Bodies Unbound

Transforming Lives Through Touch

Bodies

Transforming Lives Through Touch

Unbound

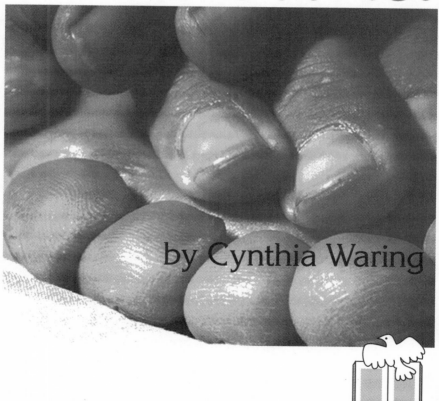

by Cynthia Waring

Books Beyond Borders
Boulder, Colorado

Published by Books Beyond Borders™
3640 Walnut Street, Suite A
Boulder, Colorado 80301

Cover and interior photographs by Sandy Corner
Book design by Rik Rydlun

First Edition
ISBN 1-883862-10-8

Library of Congress Catalog Number: 96-83288

Dedication

To my son, Russell, who said, "It stops with us."

Contents

Acknowledgments

I would like to thank my mentor, Deena Metzger, who led me, by example, to speak my truth. Special thanks also, to my fellow students in Deena's classes who encouraged and listened to my stories as they were discovered and first formed into words. Very special thanks to Michael Francis Utz who believed in me and my story. His understanding of my work and its spiritual intent was tremendously helpful. I also wish to thank my dear friends, Odette Springer, Linda Powers, Kevin Bellows, Dan Drasin, Raina Paris Ridgeway, and Jeanette Greenberg, who supported me through this arduous journey and lent me their courage and inspiration when I had none. I am grateful to those who helped me edit this book in its various forms, especially Judith Rydlun, Jacqueline Darrigrand, Michael Utz, and Sharon Weil, whose wisdom, love, and generosity have supported and nurtured me. Finally, my gratitude goes to all those brave men and women who openly and unguardedly told me their stories, letting me know I was not alone.

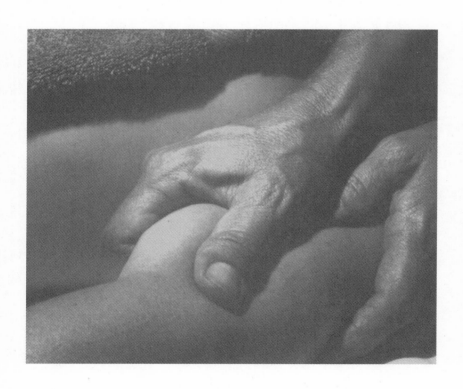

Introduction

I love bodies. I love every shape they come in. I love their smells, their texture, their heat. And I love touching them. I have been a masseuse for twenty years. The veins in my hands are like rivers. When I lay my hands on a body, just put them down on warm, soft flesh, I am in ecstasy. I love intimacy. I love knowing secrets. I love to get to the core of things. Every time my work takes a person deep into the center of her being and she remembers the beginning place of contraction and how she protected herself out of fear, I feel fulfilled.

I became a masseuse through a series of accidents in 1976, within months of leaving a convent. I went into the convent to avoid the pain of my life, to rise above it by escaping into spirit. But God knows the difference between love of Him and fear of pain, and my days in the convent were numbered. I believe God knew the way to my heart was through learning the sacredness of ordinary life and the wisdom of the flesh, and I learned this by becoming a masseuse.

I am also dyslexic. Dyslexia is a visual/brain dysfunction. It is difficult for me to interpret what I see. My understanding comes from touch, emotion, and spoken words, not letters on a printed page. I believe this is one of the reasons I felt so at home with massage. Giving a massage is a personal experience, and this is where I felt comfortable and gained confidence. I didn't know I was dyslexic when I began my career in massage. I only knew that for the first time in my life I found a way that was slow and gentle and personal enough for me to participate. At last I was needed, and in this place I began to thrive.

Because of my dyslexia, I couldn't plan my life in the ordinary way. So my life just happened. I couldn't read the newspaper to find out what was going on, so I had to depend on information coming to me from an unseen source. Perhaps this explains my profound interest in spiritual matters. I even discovered my dyslexia by accident when I was thirty-six years old, eight years after I began doing massage. I met a man at a lecture who was an ophthalmologist. I had a reoccurring eye infection that had been going on for nine years and caused me great discomfort. He noticed it and asked me to come to his office. It turned out he had written his thesis on dyslexia, and he gave me some tests just to show me how dyslexia worked. He gave me a paper to read with 100 words. I was hooked up to graphs that monitored eye movement such as how often I reread a line or how many times I fixated on a word. When I'd finished reading the 100 words, he asked me ten questions about what I'd read, and I couldn't answer one of them. I will never forget the panic I felt when he asked me a question and I didn't know the answer. It was the way I had felt in school, which had led me to think I was stupid. I remember the way the roots of my hair hurt from embarrassment and shame. I could read the words. I recognized them. But the meaning of the words wouldn't come to me, and I never knew why. I discovered through the tests that I have no short-term memory. I have to do things slowly, without anxiety, in several different ways in order to learn.

By the time I slowed down enough to know anything about what I had read, we discovered I was reading below first-grade ability. The ophthalmologist had a program that trained the eyes and mind to work together more efficiently. I participated in it for six months, and those classes changed my life. I remember after the training finding a radio schedule in a newspaper. There was a blues program a few days later. I love the blues. I remember writing down the time, the channel, and the date of the program. I even located the program on the radio dial. When all those details came together and I began listening to the blues program on the appointed day, I couldn't stop crying. I suddenly became aware of my isolation. I had been living as though on a deserted island in the middle of the ocean, having to wait and accept those things that drifted up on my shore. Because my mind could not fathom information from the printed page, the choices of events I could attend were limited to information that came to me from conversations. I now had insight into problems that had embarrassed me all my life. So that was why I always said, "Left" when I meant "Right." Or why I was always going "down" to San Francisco and "up" to Mexico. That was why I never knew about stories on the news, magazine articles, or famous people whom everyone knew about. When I told my son, he said, "So that's why you can't play Pac Man!"

After the training to remedy this disorder, I learned how to use the banks, which I had never understood. They were too complex. I liked to hand people money and have them hand it to me. It was a nuisance paying bills with cash, however, and a friend convinced me of the efficiency of checks. I had always used the pockets of my coats in the closet as my bank. When I fished out all the money from the pockets and counted it to take to the bank, I had ten thousand dollars. I had no idea I had so much. It had been collecting there for years. My clients would hand me money, and I would bring it home and put it in my coat pockets. When I needed money, I went to my closet.

Soon after that I began to write, and for the first time I discovered what I thought. My life has been about feelings and touch. It has been an awakening for me to discover the words to describe my life–words that describe the moods and stories of my life, the clients I have known and loved, the men I have known and lost, the tale of my beloved son, the child I adore. But finding these words has been a slow process for me. This book is the result of that effort. It has grown slowly, like bone and muscle.

Being focused on the present, and so connected to my clients, I have collected their stories through my hands. When I write about my clients, their memory fills me so completely that they seem to be right next to me. I have been so full of them that there has been no room for me. Their stories are clamoring inside me, crying out to be born. I am anxious to discover who I am separate from them. This, I believe, will happen after their stories are out of me and onto the page.

I have always enjoyed watching bodies tell their stories. I have spent hours at cafes just watching. There are so many lonely, sad bodies, drooped over, pelvises tilted, hidden under layers of tension. Necks that have lost their curve, flat feet, compressed chests. And then there are the runners, demanding their endorphins for the day like Catholics, their tongues hanging out at the rail, beat horses, hyped on their own adrenaline.

And there are bodies that walk like sticks; heads on top of their necks, shoulders stiff with pride, stomachs touching backbones. They do not eat. They are not nourished.

Some bodies sway and swish as if walking to the beat of Caribbean music. They melt into chairs, into doorways. They flow past parked cars like a vision.

And sometimes I see a body that bounces along, free as a let-go balloon, fresh as new leaves in spring, and I wonder, how can anybody be so joyous? How did she escape the crippling effects of our civilization? Where is her shame, her mask? What kind of person made that kind of body? What is her life like? What is her story? Through massage I have found my answers.

I have given more than fifteen thousand massages in the past twenty years, arriving with table, sheets, and oil at the homes of strangers. I've massaged every type of person, from movie stars to babies; old women dying of cancer—no breasts; men from war—no legs. I've been a masseuse to young girls as a present from their mothers like some initiation rite. I've been a present to lovers on birthdays, coming to their house with balloons tied to my bag.

I've massaged in hospitals, country clubs, beauty shops, exercise salons, and hotels and out by swimming pools getting sun stroke and in bedrooms getting scared. I've had jets sent for me by the rich and people calling for all kinds of sexual perversion.

I've massaged women who have just been raped, mothers who have lost their children. A soap opera queen cried every time I touched her; she couldn't tell me why. People have remembered every kind of emotional and physical abuse on my table—which made me remember mine. I didn't plan this trip—this journey through the bodies of earth. No, I never planned to be a masseuse.

My clients have shown me the parts of themselves they never showed anyone before. During massage, they have been open and vulnerable with me, and it was in this state of being that I was nurtured as well. My clients' need of me gave me a sense of worth I had not experienced before. My clients' trust in me allowed me to see myself as valuable. Through their eyes I saw my strengths, my courage, and my ability to love. I listened to the woes of my clients like a confessor and told them stories of my own. They let me in to see their bruises from facelifts, asked for my prayers when they couldn't get off drugs, called at 3:00 a.m. when they'd been hurt by bosses, lovers, or friends. I would go to them to help them make friends with life again.

All the ways in which a person feels shame and bewilderment and other thoughts usually kept private were common topics on my table. My clients begged me to get rid of their fat. I'd listen to the pain and grief hidden in that flesh, first with my hands then hearing their cries as the sorrow worked its way out of their bodies in the form of tears and memories. I was told about sexual problems, love affairs, and other hilarious, almost unbelievable stories.

I listened to my clients' nightmares and massaged their mothers when they came to visit. My clients would say, "Knock some consciousness into my mother, would you?" These mothers had usually never dreamed of having a massage. They were women who had never heard about "positive thinking" or "affirmations" or "reflexology." They had never heard the words "emotional abandonment" or that it was all right to grieve. Many of these mothers had never had anyone on their side before. They were shy when I said they had nice legs or hair. They cried when I touched their faces gently. They'd tell me, "Please don't tell my daughter I feel unappreciated," and I never did. These were mothers I never saw again.

Sometimes family members would give their father a massage. It was always a surprise for his birthday. The family thought Dad would love this gift, but most of the time he didn't. I could tell he felt vulnerable and was terrified he might become aroused. My touch created anxiety rather than pleasure. It was as if I were torturing him.

And when I massaged other males of our species—brothers, sons, uncles, friends—my hands on their bodies stirred up every kind of sexual dilemma possible both in them and in me.

My clients never invited me to their parties or sent me postcards on their travels. In public my clients would hide. They hid from me in grocery stores, their carts filled with puddings, cream, and whiskey. I would turn my head so I would not embarrass them. I didn't care what they put in their mouths. I just wanted them to love themselves. I wanted them to be able to say in the mirror and to all their friends, "I ate every bite, and what's more, I loved it."

The life of a masseuse was not always pleasant, deep and touching. Many people sneered when I said I did massage. I could tell they thought, "Whore" or "Servant." Clients would call at the last minute and be outraged if I couldn't come. Sometimes I'd drive half an hour to get to a client, and then he wouldn't be there. I had no insurance and little respect. Wealthy people often quibbled about $5.

The following were common phrases I came to expect: "Is your oil natural? Don't ever touch my stomach. I want it upstairs–can't you carry your table upstairs? Don't get oil in my hair. I hate it when you rock me. Harder. Lighter. Massage me while I'm on the phone. I have to catch the news. I still have that knot! Can't you come later; I want to go to bed after my massage? If I'm not here, just wait. Don't mind the dog. Change the music. Take a message." But no matter how difficult a person was, I usually found the loving part of her, or perhaps I should say, massage found the loving part.

Being a massage therapist changed me as much as it changed the people I worked on. Through this profession I came to have a confidence and an identity I had never had before. At first I discovered my immense need to be needed and to please others at all costs. Both men and women brought every kind of temptation to my life, and I fell into one pit after another. It took a long time before I could say, "No, I will not come to your house at 4:00 a.m. No, I will not wait for you for two hours without getting paid. No, I will not lug my table up two flights of stairs in the home of a weightlifter. No, I will not touch you there."

Finding the confidence and self-love to say, "No" to those demands was like going to the bottom of the ocean to dive for pearls. And how I found them, those words, those pearls, ah, now there's a story.

Signs

If someone had asked me what I wanted to be, I never would have thought of being a masseuse. I simply needed a job. It was hot. My feet hurt.

Walking through a shopping mall, looking for work, I saw a sign in a health spa window. The sign read, "Masseuse Wanted." I thought, "I bet I could do that," and I went inside and applied for the job. The manager was flirting with some guy. She said I would have to come back the next day to massage the owner. She didn't ask if I'd been to massage school, if I had a license, or even if I'd done massage before. Thank goodness she didn't because I had never had a massage, much less given one.

Going home on the bus, I saw another sign. It read, "MASSAGE" in great big letters. I got out at the next corner and walked back to the massage office. I planned to ask someone there to show me how to give a massage. I was just about to ask the masseur on duty when it occurred to me that he might take a dim view of someone right off the street asking him to teach her his profession. Standing there, needing to say something, I asked him how much he charged. When he told me a massage was $35, I said, "I'm sorry, but I don't have enough money." Plus I was not prepared to take my clothes off in front of this stranger, no matter what the cost. I was turning to go when, to my surprise, he said he wasn't busy and he'd give me one for free. Not knowing how to say, "No" to this offer, I suspiciously took a shower as he instructed. With great trepidation, I lay down on his table. While I was lying there waiting for him to come into the room, I realized I could memorize what he did. Then I would know how to massage the owner of the health spa the next day. Very pleased with this chain of events, I waited for my massage to begin.

Just a few months before, I had left the convent after a year. The sisters didn't believe in touching. Prior to that I had spent years and years at spiritual pursuits trying to find a way to control my naturally sensuous flesh. I had stared at blank walls following my breath for hours. I had chanted myself into a stupor. I had prayed on my knees on hard wooden floors, eating nothing but brown rice and miso soup, hoping to purge myself of all my demons. It had been so long since I'd been touched that I felt as if I'd never been touched before.

When the masseur came in he turned on a tape player. Music from heaven filled the room. He began the massage with my feet. I heard angels singing. He opened the dammed-up rivers in my thighs and calves. I sighed, letting go of tension. My arms came alive in his hands. I felt blessed when he touched my face. My back was restored as waves and waves of feeling surged out my arms, down my legs, up into my face, into my belly, around my head, and splashed over the boulders in my back. Finally I knew that on this earth I was my body and on this table I had been called home.

I was in awe. I didn't know the body could feel such pleasure without feeling guilt. Not only were my knots being broken up, but emotional armor was crumbling too. I felt whole, ecstatic, integrated, loved, and calm, all at the same time. After all my searching for a god from all the gurus, in all the ashrams, churches, religious orders and drugs, I had never felt so holy as I did on that man's table. I felt accepted and cared for, even loved.

I, like everyone else I knew at the time, believed the body could lead me astray with its feelings and desires. I had tried to stifle my body, to make it feel what it was "supposed" to feel, not honor how it truly felt. I don't know how many times I'd heard, "You shouldn't feel like that," or "You should feel grateful" or "ashamed" or anything but what I was truly feeling.

I had fought against this training, but the punishments our society puts on a girl enjoying her body are immense. I had suffered tremendous guilt for my sexual desire. When my mother learned I had discovered sexual pleasure via masturbation, she gave me a book written by a doctor that said if a girl masturbated she would go insane and never be able to have a relationship with a man. Mom and that book scared me to death.

A life of passion, of following the body, the physical manifestation of the soul, I have found, is a difficult life. I didn't follow the rules. Consequently, I didn't know anyone with similar life experiences. Often my life was such a mess that I didn't know where to turn or how. But my intuition would always guide my way. Without any apparent guidance at all, I began to trust that my life would turn out all right. I came to depend on my intuition.

Then, as if my own experiences weren't hard enough to bear, there was every form of judgment against such a life. There was Christian religion with all its taboos, and Indian philosophy saying, "All manifestations are an illusion; rise above everything you can see, taste, touch and smell." These ideas perpetuate the message that thought, or even nonexistence, is a higher form of being and that somehow we are supposed to ignore our bodies. But if life is the Father, isn't the body the Son?

On the masseur's table I realized that as a human being I could not separate my experience from my body. On earth I am my body. I am also my body's environment.

On that table I became a priestess, an ancient one who could lead her people into rites of passage and rituals uniting them with the wisdom of the body. At last I knew what I wanted to do with my life. I wanted to anoint all my brothers and sisters with oil. I wanted to bring their minds and souls and breaths of life back into their bodies, their earth. I wanted to witness their sighs when they came home. I wanted to roll away their blocks of shame, to break up their knots of guilt and despair, to unite them with their senses and passion–their nature–to make them one with the honor and wisdom of that nature.

I knew I wanted to be a masseuse when I left the office of that wonderful man whose name I never knew. I went out on the street, alone once more, waiting for the next bus, feeling confident. I had an experience of well-being. I knew I was going to be all right. I had found a place for me and a purpose.

The River, The Fire, My Son

I had lost all confidence in my own feelings, in my ability to know what to do, when I saw the sign "MASSEUSE WANTED." I didn't know what I thought. I had lost everything that was dear to me. I had walked away from what I loved the most, my son, Russell.

Phil was my first husband and Russell's father. Phil and I met in 1966 at San Jose State. We could not have been more different. Phil had started the underground newspaper there and was at the hub of student politics. I barely knew Reagan was governor. The noises in my own head screamed louder than anything outside myself.

Phil brought me a new world that seemed to matter. He brought me the world of Vietnam. He brought me Africa, civil rights, Birmingham, injustices from around the globe, and the plight of people who looked as I felt. I was so overwhelmed with gratitude for finding some form of life I could sink my teeth into, something real and visceral, something important, that I couldn't separate this feeling from the source, Phil. I admired him tremendously. It took me a long time to see that "political" people tend to love others abstractly; they don't love their smell or stare deeply into their eyes or embrace their weaknesses.

Oh, and how my parents hated Phil. He was a "communist," a hippie. He had a beard and wore sandals with socks and Bermuda shorts. His hair went beyond his ears, he wore John Lennon glasses, and he was against the war in Vietnam. And Phil, with his brilliant, Jewish, photographic memory, knew every law the U.S. government was breaking to go to war. And he brought all this up at dinner in front of my parents' Protestant, Republican, wave-the-flag ideals. My parents could not have been more horrified if Phil had been black and deformed.

I became pregnant after being with Phil four months. He said he wanted to marry me, so we moved in together right there on campus. My parents disowned me, but I got caught up in the excitement of it all. The night before our wedding I admitted to some girlfriends that I really didn't want to get married. I was scared. I didn't think I loved Phil. They made me call Phil and tell him. We cried all night.

The next morning I kept the appointment I'd made at the beauty parlor, and there, under the dryer, I began to miscarry. Phil was called, and he took me to a hospital. There they gave me drugs to prevent the miscarriage and sent me home. Friends came over to see me. They brought the school newspaper, which ran the story about Phil and me getting married in the school chapel, and that the president of the school was coming. How could I get out of the wedding? It was right there in the paper.

With the drugs, my miscarriage slowed down. I called my history teacher and asked him to walk me down the aisle. Phil's best man was wanted by the police for blowing up a building along with some other SDS members, so we had a lookout posted for the Feds; another one for my father, who had threatened to stop the wedding; and another for an old boyfriend of mine who had just arrived home from basic training and was threatening to abduct me. Phil's mother,

Sylvia, was crying her eyes out and refusing to speak to anyone because Phil wouldn't let her walk him down the aisle in the old Jewish tradition. I was having contractions and was stoned on codeine.

After the reception Phil and I drove to Carmel for our honeymoon. On the way I began bleeding heavily, and we ended up at a hospital, where I had a D&C. The baby was lost, but there I was–married.

Being married to Phil was an adventure. A few months after the D&C we went to Europe and hitchhiked from Madrid to Istanbul. When we got back we settled in Cambridge, Massachusetts, where Phil went to Harvard, and I got a job to pay the rent.

The marriage looked idyllic and it made a good story, but I was unhappy. I did not know how to love. Being close to someone terrified me, and in marriage there is no escape. I ended up coping with my terror by misusing drugs and alcohol, which had been my friends since high school. Then I got pregnant again. I didn't want to be pregnant. I didn't want to be married. I looked in the paper every day at the "For Rent" section, planning and dreaming of a life of independence from responsibilities, of romance and adventure. I didn't want to be a secretary. It didn't matter that I worked in William James Hall, where I met Erik Erikson and Professor Jerome Kagan. I felt stupid around all those intellectuals.

I was not in love with Phil until the night our son was born, when Phil rubbed my back during contractions and hid himself in the food cart outside the delivery room so he could be close to me in case I needed him. My heart cracked open with the birth of my son, and then I loved his father.

I gave birth to my son, Russell, named after Bertrand Russell, on November 3, 1968, at 8:05 A.M. in the Cambridge City Hospital after a long labor. I loved him from the moment I saw him.

While he was still a baby, I would wake him from a sound sleep just to hold him. I could stare at him for hours, just watching him breathe. I thought up songs for every occasion: songs for waking up, songs for going to sleep, songs for putting on shoes and socks, and songs for taking a bath. I nursed him for his first eight months. I became obsessed with health food. I couldn't stand the thought of anything remotely resembling a chemical entering his or my body.

I could not keep this up, however. After a year of trying to be the perfect mother, I became restless and harried. I broke my kneecap ice-skating and had to manage my fourteen-month-old son alone while in a cast from ankle to thigh. It was winter and I couldn't go out for fear of slipping on the ice. I felt that Phil and I lived separate lives, in two rooms close to Harvard College. Phil was on campus, either studying or leading rallies against the war. He didn't know I said good-bye each morning with a smile induced by speed, greeted him each evening in an alcoholic stupor, and got to sleep by smoking hashish. Sometimes I wondered what they taught there at Harvard. I felt so alone and so scared. I was terrified I wouldn't do the right thing for my son, terrified I couldn't resist the pleasures that kept attracting me.

Oh, the things that took me away from my baby, the things that deprived me of being a mother. The lure of the perfect high from acid, gurus, or miracles. My old college roommate, Judy, now a flight attendant, flew in once a week to Boston with the best drugs the 1960s ever made. It was the year of Woodstock, the year America marched on Washington, and I was home washing diapers after countless shots of Ouzo and myriad tokes of grass, waiting for a husband who was out doing exactly what he wanted to do.

I tried to be a good mother, but it seemed the only way to be a mother at all was by taking enough drugs to keep away the anxiety. Ultimately, drugs scrambled the messages from my subconscious that would have warned me of an impending disaster.

One day while my old college roommate was visiting, we took acid together and hitchhiked out of town for the day, leaving my son with a baby-sitter. At dusk we realized how late it had gotten, and I was desperate to get back home. The acid was wearing off. Reality was setting in. I began to re-enter a world where being responsible was important. I began to notice the broken beer bottles in the park, the used condoms, the oil spills in the parking lot. We saw a couple kissing in a nearby car, and we asked them if they could take us back to the city. And he said, "Why not?"

Had we looked at them closely, with clarity, we would never have gotten into the car. The man in the driver's seat was huge, well over three-hundred pounds. When he smiled, I noticed his teeth were discolored and rotting. When the woman turned around to greet us, we noticed her eyes weren't on the same plane in her face. She looked like a Picasso. She had her long black hair in a tight ponytail. Her lipstick was smeared all over her face from kissing.

When we were arranged in the backseat and the doors were firmly closed, the driver screeched out of the park like someone mad and haunted. As we barreled down back roads, over hills, and around curves, just missing parked cars by inches, Judy and I held on for dear life. Our eyes were as large as golf balls. The man never slowed down as we neared the city, still off the main streets. We entered a one-way street going the wrong way. A sign that clearly said, "DO NOT ENTER, KEEP OUT" didn't faze him a bit or slow him down. We crashed through the sign, drove on the sidewalk and in the road, where there were huge holes in the pavement made by bulldozers. We crashed through sawhorses supporting warning signs, throwing them everywhere. We drove across lawns and over curbs and finally onto a main street right in the center of Cambridge, where we screeched to a halt at Harvard Square. "This okay?" the driver asked. Picasso turned around and smiled sweetly, exposing her toothless gums. Much later I realized the ride represented the life I was leading. Picasso and her old man were creatures lurking in my psyche, driving me to create one crazy experience after another. They were the parts of me I did not know, who stole from me every form of beauty and blocked out every bit of grace.

But at the time I didn't know about the shadow side and my drinking got worse, further impairing my judgment. Three incidents convinced me to leave my son. In the first incident we were playing on a beach. Russell wanted to go into the ocean in a yellow rubber boat some friends of ours had brought. I put Russell in the boat and pushed him out through the waves. I was swimming alongside, pushing him farther and farther out, when a huge wave appeared out of nowhere and crashed over us. When I came to the surface and looked around I could not see the boat or my child. I began diving into the murky water, panic flooding my body. I was hysterical, diving and coming up for air. I kept thinking, "He's underwater and I can't find him. How long can he make it without air?" Someone finally swam out to tell me Russell was on the beach. His naturally buoyant, two-year-old body had been deposited on the shore by the wave. I shook for days. It was as if my body was trying to shake me into some awareness.

Several days later the second incident happened. While driving home from the ocean, I ran a stop sign and was broadsided by a car that had the right-of-way. I didn't run the stop sign deliberately; I just didn't see it or the car or know I was driving at all. Russell and I were not hurt, but I became terrified, and Phil was beside himself.

The last event took place at a wedding for some friends of mine. I got very drunk, then asked for some speed so I could stay awake to drive home. I remember putting Russell in his car seat and backing out of the driveway. That was the last thing I remembered for the next two hours. I came to my senses just in time to recognize the turnoff to our street. I had driven two hours in a complete blackout with my little son in the car. When I became conscious, Russell was staring at me. I didn't know if he was trying to speak to me or trying to reach in and find his mommy. I don't know why he stared at me. I can only imagine and remember the times I stared at my parents, bewildered, wondering, "Are you there? Is anyone home? Am I here or am I invisible?"

Phil and I were divorced in 1970. He wanted custody of Russell and I agreed. I really didn't know what that meant. I didn't know how my life had been defined and protected by the needs of my son and my husband. Russell was almost three; I was twenty-four. I was so young that I didn't know the decision to leave Russell would cause me so much pain. I didn't know the human heart could suffer so much without dying. Perhaps it was this need for a family, this need to mother, that drew me to massage therapy.

At the time I left Phil and Russell, I knew leaving would be hard, but it was much harder. I walked and cried and drank for months. It was worse than I could have imagined as I crawled out of my despair and my flesh awakened from the shock of misuse. I was jolted out of the numbness that I was experiencing when the realization that I had truly left my son sank in. I knew then that I'd made a terrible mistake.

I had no home. I was on the streets and sleeping in abandoned buildings with people who had also abandoned all hope or struggle. I felt as if I were on an endless walk in the desert discovering the deceit and falsity behind every illusion. I reached a state of hopeless degradation and despair and wore out every possibility of escape.

Each morning I awoke thinking I was still with my son and husband. I could hear the sounds my son made, the cooing as he woke from a nap or a night's sleep. "Mommy, Mommy," he would say like a cat purring, like a brook murmuring.

When I realized the truth of where I was, on an uncovered mattress in an abandoned building where I ate food from cans, I would feel a stab in my heart. When I knew that I was dreaming, that my son was not really there, the pain spread from my heart down my arms, filled my stomach, and raced down my spine. "He is not here," the pain told me. "Your son is not here. You left him."

And the emptiness over the loss of my son would not be filled. I caved in around that black, empty space. It pulled me in head first.

The numbness began to wear off one day while I was drinking tea in a health food restaurant in Harvard Square, where I made salads and washed dishes. It was a tea made from reeds found in the desert. As I put the cup to my lips, I knew that neither the tea nor anything in this world could touch or comfort me. I had left something behind that I could not release. I became a shrine to the loss of my son.

Phoning old friends did not help. Rushing into the arms of men did not help. Smoking grass did not help. Nothing helped. My son's bewildered face would rise before me, float above me like a moon I could not reach. He came through the haze of every binge. There was no thought or dream he did not enter. It was as if he were the dream, the sleep, the mist, the hole, the star, the moon, the grief, the abyss, the nightmare, the dinner, the prayer, the God, the beating of my heart, the razor, the wrist, the blood, the blight, the fog, the rain, the mud, the sun, the pain, the silence, the torment.

I drank the tea and silently screamed and drank the tea and silently screamed. The tea made from reeds found in the desert–long, thin, green reeds that separated into sections so easily, not like my son and me, no, not like us.

Oh, yes, I was free now, free to get raped while

hitchhiking, free to cry myself to sleep every night thinking of my son. I was free to drink myself into a stupor, free to take enough drugs to become an addict, and free to wake up with major anxiety attacks. I was free to do all of these things over and over and over again, until some blessed inspiration burst forth in my brain.

The man I was living with had an affair with a friend of ours who came to visit for the weekend and wouldn't leave. I thought to myself, "I won't kill myself. I'll join a convent."

Joining a convent was not an out-of-the-blue decision. I had been prone to mystical experiences my whole life. I had been involved with yoga, Sufi dancing, Indian gurus, Christian mysticism, palmists, astrologers, I Ching, runes, tarot, sweat lodges and rituals of all kinds. I hoped the convent would help me escape the mess I'd made of my life and give me direction. I was grateful to find a place where I could rest and recover.

This tendency to go off the spiritual deep end looking for a savior was balanced by my obsession with sex. I would swing back and forth between sex and spiritualism, experiencing a broad variety of possible lifestyles. I would be a saint for a time. Then I would hit a certain degree of safety or boredom and toss all my hard work into the air. I would break the spiritual part of the cycle by doing something so outrageous that it would take several months for my own psyche to comprehend what had happened, let alone friends and family. This would lead to feelings of degradation, which would eventually lead me to pious acts of penance.

After I had been in the convent a week, I began to detoxify from a gallon of wine and a pack of cigarettes a day and all the drugs I had been on for years. I never told the sisters why I was so sick. I think they thought all the evil of the world was coming out of my mouth, my bowels and my pores. It's as good an explanation as any, I suppose.

After a year I became restless again, and I knew I had to leave. In the Celtic religion my name, Cynthia, is the name of the Goddess of the Moon, and like the moon, I am always changing. The phases of the moon are present in me. I do not control which side faces where or how much of me faces the sun. I was forty-five years old before I learned to have faith in those moments of darkness and change, to simply endure.

The day I left the convent I was cooking lunch for the twenty women living there. I was also high from drinking sherry, which I should have been using to marinate the roast. A beautiful man came to the back door selling fish. He asked, "What are you doing here?" I answered, "I don't know."

Within minutes I had packed my tiny bag of belongings and was riding down the road next to him in his beat-up old pickup truck, which reeked of halibut. There was a song on the radio by a group named Hot Chocolate. Ricardo sang to me at the top of his lungs, glancing at me seductively while driving at breakneck speed into the fading light of day. He sang, "I believe in miracles since you came along, you sexy thing."

I had been aware for two months that I would not be able to stay in the convent much longer. The realization came to me when I read a letter from my mother about my son visiting her. She wrote that she had walked into her bedroom and found Russell holding my picture, kissing my image through the glass. I could imagine his seven-year-old hands holding the only thing he could find of me. I was sure there was not a trace of me to be found in his own house. The only thing his father and stepmother allowed him were my letters and my down sleeping bag, which I had left him when I went into the convent. Russell and I called this sleeping bag "Blue Cloud." He said he liked to sleep in it, but he didn't use it often because his stepmother made him stuff it back in the bag each morning all by himself and that was very hard to do.

I could not remove these images from my mind: my son kissing my picture and stuffing Blue Cloud along with all his feelings of mother back into a small bag in the morning. These images yelled at me, "Wake up, wake up." So when the fisherman came to the back door of the convent, I knew he was my ferryman.

I moved back home with my mother and stepfather in their walnut orchard in the San Joaquin valley, where I had been raised. I sat in front of the fireplace on cold, foggy December nights and watched the prunings from the walnut trees burn and turn to ash. I stared at the fire for hours, feeling at home with the flames. My parents stared at me, then at each other, afraid of the emptiness I had become.

There was a dirt road that went beside our walnut orchard and through fifty acres of asparagus behind the trees. This road led to a river, where I sat on a footbridge made of rope to watch the water. I walked there each morning at dawn through fog that came all the way down to the ground. I felt wrapped in a shroud by the damp softness and surrounded by something ethereal, impermanent, ghost like. Sometimes, as I walked down the road, I played a bamboo flute my mother had made as a young girl. One morning Don, who owned the asparagus field, was checking his levees when I came down the road playing the flute. I nearly scared him to death. When he finally saw me, I could tell he was badly shaken. He said he kept hearing this eerie sound, but all he could see was the fog. He thought he was losing his mind.

When I went home to the trees of my childhood, I realized what an alien creature I had become. Living in strange places, I couldn't tell who I'd turned into. I had to go back home and measure myself. The things I used to walk under now hit me in the chest. All the houses looked smaller, the long dirt road looked shorter, the faces of the neighbors were more wise or more shallow.

My son came to visit me that Christmas. I hadn't seen him for a year because I had been in the convent. At the airport we were both struck by the force of our emotions. Neither of us spoke. Thank goodness there were people around us to make some noise and help us remember words like "hello" and questions like "How was your trip?" I had held him in my heart day after day, trying to bring back the way he looked, to make some connection with a picture I took out of a secret place; now the real person was too much for my mind to comprehend. I was overwhelmed with fear at the sight of my son, and I could tell he was terrified of me.

I did not feel safe with Russell until he had closed those gigantic brown eyes and was sound asleep in the soft warm bed of my childhood. I stared at him by candlelight and once again watched him breathe. I sat beside him like I sat beside the fire and the river, just watching him. While I sat there something began to stir in me. Something began to touch me. It was the need Russell had for me that called me out of my lethargy. The fire and water could engage me, but they did not need me.

The way my son loved me, and the way he patted my face felt primal, connected to the earth. There was something warm that reached to the core of me and let me know there was no one else in the world who could take my place for him. This knowledge came as a shock and a blessing. It gave me someone to be, a mother.

Ten days later, when I borrowed Mother's car to take Russell back to the airport in San Francisco, I thought I would die from grief. I watched him walk onto the plane, his eyes full of tears. Panic flushed his face and my heart as I watched the feelings he couldn't hide capture him and hold him prisoner. I could do nothing. I could not hold onto the one I loved; his father had custody of him. I couldn't save him or myself from these feelings; I could only watch. The world felt like one large prison.

We came together and separated, came together and separated, for eight more years, growing together and ripping apart. It was a slow, tortuous growth.

After I left him at the airport, I drove to Sausalito and walked the streets trying to connect with what I saw and felt. There I met two men, and the three of us spent the evening dancing. I remember very little about them except that they were not from the United States, and they were kind. I danced first with one, then the other. Their arms held me together until I grew new skin and my bones hardened. I danced with these men until there was enough of me back in my body to safely drive home.

When I got home, I realized I could not stay in my mother's house or in that community any longer. A few days later I packed my things and moved to Los Angeles.

I moved with two goals in mind: to create a life that supported freedom, and to make a pleasant, safe home in which my son could come to visit me.

So when I saw the sign that said "MASSEUSE WANTED" and something inside me said, "I can do that," it wasn't out of arrogance or an unshakable belief in myself. It was more like a miracle. Life had opened herself up and said, "Here, we found a place just for you."

My First Massage

Whenever I remember going back to massage the owner of the health spa I wonder, "How did I do it?" I walked into the office and met the owner, who led me to a back room, where she took off all her clothes right in front of me. She handed me a bottle of oil and lay down on the table, covering herself up with a towel.

Before the massage began I figured, "There was nothing to it. Rub a little oil into someone's skin, move your hands in various ways and pronounce them massaged." Now I viewed massage as a holy ritual.

Standing there before the owner of the health spa, the one who expected me to touch her expertly at any moment, I froze. I went completely and totally blank. I couldn't remember a thing the masseur had done on me the day before. Fear raced through my arms, legs and heart. Then I heard a voice in my head give instructions. They were not "Build an ark" or "Change this water into wine." They were "Do something. Pour oil in your hands. Put your hands on her foot and start rubbing."

I was aware of my heart thumping as I poured oil onto the palm of my cold hand. It took every ounce of courage I had to put my cold, shaking hands on her foot.

I realized an amazing fact as soon as I touched her. When I put pressure on her body with my hands, they stopped shaking. The warmth of her body warmed my hands.

I was glad she had her eyes closed because I poured too much oil on my hands and had to wipe the excess off on my arms. And, with one hand on her foot, massaging all the while, I leaned over and put more of the extra oil on my legs. I felt like an acrobat in a circus. As I worked my way up her leg, I wondered how far it was proper to go. What if I touched something I shouldn't? The towel was in the way. I couldn't see where I was going. What did I do with the towel? Any minute now I was sure she was going to open her eyes and say, "This is the worst massage I've ever gotten. You haven't the faintest idea what you're doing."

Sweat poured off me in the hot room; maybe I would faint. I kept looking at the clock, wondering how long to give each part. Ten minutes had passed, and I was only up to her knee. How did I get from her hip to her stomach? Did I take the towel down from the top, uncovering her breasts to do her stomach, or lift it up from the bottom, exposing her pelvis? Either way embarrassed me. Did I do both legs first or go around her in a circle? Did I massage her arms while she lay on her back or while she lay on her front?

As I neared her head, forty-five minutes later, the idea of touching her face made me cold. I don't believe I had ever contemplated the intimacy of a face. Not only that, she hadn't said a word. She just lay there. I didn't know if she liked what I was doing or not. Was she even alive? What was she thinking?

When I asked her to turn over, my voice shook. I waited a full minute to get the nerve to speak. Had I done enough on her front? Was I enough? All I knew for sure was that she was oily. I wondered what she'd do if I left. I could just walk out the door. She would never know where to find me.

When the owner of the spa turned over, I found I was more relaxed with her on her stomach. Her face was down. She couldn't open her eyes and catch me staring in confusion at her body, wondering what to touch next.

There was no way I was going to finish the massage in an hour. With ten minutes left, I had half a back, an arm, and two legs to go. I said, "I'm running late." She said, "Take your time."

Her voice was dreamy, like she was enjoying it–like she never wanted me to stop. A trickle of confidence seeped into my veins. Take my time? What a concept. They'd never said that when I was a secretary.

And with the confidence that followed, I remembered the way I had felt the day before when the masseur had massaged me. I concentrated on my touch. It was my voice. I was singing to her hips, opening them to freedom, unlocking the doors, stretching, soothing, caressing. It was my song, like a brook babbling; swishing through the rocks, over, under, around and down. I swished through the curve of her back and into the knots in her neck and over her buttocks and down her legs; going slowly, backing up, faster, then ebbing and stopping at a place, holding, pressing, then letting go– and my song was over.

I don't know if they needed a masseuse so badly they would have hired anyone with two hands who said, "Yes" or whether she truly liked my massage. Whatever the reason, she said I would do, and that's how I became a masseuse.

Getting Started

Most of the women who came to the spa where I had my first job were elderly. I wandered around in a white uniform trying to look helpful. I gave them a hand getting out of Jacuzzis, stinking of chlorine, and found them extra towels. When someone wanted a massage, I was paged over a loud speaker.

Almost everyone I rubbed had suggestions: their last masseuse had put the towel just so, heated their oil, cracked their neck. My massage was "simple," they would say.

But what I lacked in knowledge about strokes and anatomy I made up for in sympathy. I could listen to and understand anything. My mis-spent youth was finally coming in handy. I understood their hangovers, their headaches, their bad marriages, their desire for a new religion, their need for my prayers, their desire to lose weight or gain weight, their worries about their children. I would listen to their stories and tell them stories of my own that would inspire another story from them, and on and on we would go until we knew each other like old friends.

While working at the spa, I never felt as though I took a mind or soul to a deep realization of their inner truth. I was too shy, too unsure of myself, too nervous. It was later that I realized I couldn't go any more deeply with someone than I could go with myself.

After I had been a masseuse for three months, I decided to get some training. I wanted to learn a routine. I wanted to know the names of the bones and the muscles. I wanted some confidence. Not knowing the names of the body brought up old fears of not knowing something I was expected to know. When the fear grew intolerable, I called three massage schools, picked the cheapest one, and two days later became a student.

This was not an expensive, New Age massage school where they taught every style of massage. Nor was it a school that played music of birds chirping by a waterfall or raindrops and wave music. I went where they trained women working in massage parlors. I didn't know that when I signed up, but it became obvious very soon.

The massage parlor ladies, most of whom were Asian, were tiny people with long black hair and soft, bell-like voices who chattered like brooks. Their gaiety impressed me; their lives shocked me. They survived by catering to men's lust, men who were dying to be touched, who only wanted sexual pleasures. The instructor would show them a Swedish routine, but the moment his back was turned, I would see them in the corner walking on each other's backs, light as leaves, balancing along the spine, laughing and talking a mile a minute. I'll never forget one woman coming out with, "I keep saying, 'No hunka punka, no hunka punka,' but he no listen." All the other women in the class laughed as she pantomimed his prick under the sheet.

I was curious about the massage parlor ladies but never wanted to be one. I was going to be a healer, a savior. Everything nice, safe and respectable. No hunka punka on my table. I would give the word "massage" a respectable meaning.

I faithfully attended classes three times a week, learning

that shoulder blades were called scapulas, rubbing was called effleurage, kneading was called petresage, and lightly hitting the body was called tappotment. At the end of two months I took the test given by the state that asked for the name and function of every major bone and muscle. Then I gave the instructor a massage and graduated.

Two weeks later I received my diploma in the mail and with it the confidence to go into business for myself. I had cards printed and drove around town putting them up on every public display board I could find. I also put an ad in a newspaper for the first and last time and my name in the phone book under "massage." As a result, for the next four years I had calls at all hours of the night. These callers panted lewd suggestions in my ear.

After all my effort I had three massages the first Saturday I was in business. My first client was a man who wanted me to come to his home in the Santa Monica hills. I got a map and found my way. I was nervous. I wasn't sure this man believed me when I told him I did "therapeutic" massage. What if he wanted sex? He was a stranger. He would be nude. It is one thing massaging a naked friend under a blanket or a fellow student or someone in a spa. Doing "out-calls" is a different story.

When I found the house where I was to give my first massage as an independent, official, licensed and certified masseuse, the driveway was so steep there was no way I could drive up it or carry my massage table. I walked to the door, bottle of oil and towel in my hand and knocked.

After considering the situation, we decided not to haul the table up the driveway. We agreed I would massage him on his bed. Inside I thought of my beautiful new table with the bridge suspension legs sitting mute in the backseat of my car.

The man I massaged moaned throughout the whole

treatment. Every time I massaged his leg, his penis, out of sight beneath the terrycloth blanket my mother had hemmed by hand, would flap around like a fish dying on the shore.

He paid me $25, and I was off to my second massage. This man worked on an oil rig and was in town for the weekend. He was huge and very annoyed because I didn't want him to massage me. My third client was a man who never stopped smiling at me seductively, his gold teeth glinting in the half-light.

I came home with $75 and a leaden heart. I had a vision of what I would be for people. I wanted to bring love to my clients, not sex. Comfort, not lust. It was a very old pain. All I had ever wanted was to inspire goodness in others. But I had never known a man, not even my father, who had not desired me sexually. Instead of getting angry, I blamed myself. What was wrong with me? What energy did I put out that attracted this? It was thoughts like these that made my heart ache.

Another man called in response to my ad in the phone book. He called around dinnertime once or twice a week to ask if I'd beat him. We'd chat and I'd say, "No, I don't do that." Once he said I seemed so nice that he had decided to try a regular massage, but I refused.

I went out every day to let people know about my business. I went to beauty parlors and left my card. I talked to hairdressers and offered them a cheaper rate for every client they found me. I talked to waitresses and offered them discounts. I went to all the major hotels in town. They all started sending me business. But none of it was steady.

One day while I was out trying to drum up business, I went to a country club. The club offered me a room for free to massage its members. The room was in the basement, next to the lockers, which banged and clattered every time someone used one and ruined the peaceful massage I was giving. It was a horrible space, drafty, noisy and cold. A month later the pipes broke, and I did a massage sloshing around in water up to my ankles. I said, "I've had it," and walked out, informing the manager that I was quitting.

That night I was so angry that I demanded from the universe a clientele made up of women in a place that was warm and dry. If these demands weren't met, I would do something else.

Two days later I got a call from Ted, the owner of a beauty parlor where I had left my card. His regular masseuse was sick, and he asked if I would like to fill in for a while. It was one of the most elite salons in Beverly Hills and I jumped at the chance to work there.

Ted's masseuse never did come back. So I began working four days a week at the beauty parlor and two afternoons a week at the health spa where I had gotten my first job as a massage therapist. In between appointments in those places I would race to someone's home for an "out-call" or to my house, where I had set up a room just for massage.

I was not an instant hit at the beauty parlor. All the women who came to the shop loved their old masseuse. They simply did not want to try me. "I didn't look strong enough, I was too young, I was too thin." So the owner put an ad in the paper for cellulite massage to draw new people.

I read volumes about cellulite. I Xeroxed a picture of a body to write measurements on and studied what one should eat to lose weight. I began beating people's bottoms and thighs to a pulp. Cellulite massage was not my idea of bringing the mind and soul and breath of life into the body, but it was paying the rent.

I asked the ladies from the country club to come to me at the beauty parlor, and when the beauty parlor ladies saw I had honest-to-goodness clients, they began using me too. I soon had more work than I could possibly handle.

A woman named Joan came to me in the beauty parlor for a regular massage and for some advice about losing weight. I gave her the first massage she had ever had. After my treatment she was so relaxed that she had trouble driving. Fearing for her safety, she asked if next time I would massage her at her home.

So every Wednesday night at eight I arrived at the entrance to her estate and rang the bell from my car. The gate would fly open, and I would enter a whole new world. Statues lined the driveway, banners flew from flagpoles, flowers grew everywhere, and a Jacuzzi gurgled next to the large swimming pool. Joan's very thin husband always met me at the door and insisted on carrying my table up the stairs, for which I was grateful.

It was a shock for me to be in the home of a truly wealthy person. Joan had a closet just for shoes. The heels hung on the wall, suspended from wooden strips to keep them from falling. They were all arranged according to color.

I never knew a person with so many shoes. Joan had never known a person with so many strange ideas. We were fascinated with each other from the start. I was a New Age massage therapist in southern California in the 1970s, and I was into everything from mucusless diets to wheatgrass enemas.

I don't know if it was her work as a writer or her inherent curiosity, but through Joan I discovered what a storyteller I was. She could pull stories out of me quicker than anyone I'd ever met. In fact, before meeting Joan, I didn't have the faintest idea I knew so much. In no time she knew everything I knew about pressure points in the feet and their relation to all the organs in the body. She knew what I thought about religious cults and who joined them, my personal stories about Sai Baba, Kirpal Singh, and Ram Dass; what I had read about food combining, lecithin and garlic capsules; my life as a hippie; how breast feeding develops both sides of a baby's brain; how to heal cancer through macrobiotics; colonics and so forth.

After six months of massage Joan began to open up herself. She had lost at least forty pounds as a result of my pounding and her restraint, and the marvels of lecithin, garlic, and lots of water, and then she began to break down emotionally.

Through Joan I saw, for the first time, how unresolved conflict is expressed in the body. I learned how the body acts out our core beliefs. And when we go against these beliefs, there is hell to pay. Underneath all that weight and tension lived Joan's fear of her softness, vulnerability and feminine power.

I could not have been more pleased for her emotional release. I had finally found someone who would work with me in the way I wanted to work. Joan, on the other hand, thought she was losing her mind. I would walk into her beautiful bedroom and set up my table, then she would lie down and sob. What came up was her mother. Joan's mother had trained her to honor her intelligence and somehow relayed that it was cheap to have an attractive body. When Joan lost all that weight and became a beauty, the war began. One minute she was raging about the stupid ideas her mother had planted in her; the next she was grieving about all the ways she had been trained to hate herself.

Watching her struggle through her problems with "mother" brought up my own issues. My mother was afraid of her own beauty and of attracting attention. She did not know how to keep my father (not that anyone could), who was a restless, sensual man at home. Someone would invariably tell her they had seen my father with a woman they thought at first was her. Sometimes husbands would come to the house asking Mom if Dad had been out on such and such a night.

Mom would be resentful, moody and silent. She would go for days without talking while Dad was having an affair. Another sort of woman might have gone to the other woman, scratched out her eyes and danced for Dad until he wanted her again. Not Mom. She unconsciously sent me. I would have done anything to make Mom happy.

Later Mom would be jealous of my closeness with Dad. She was always forcing her will on us in any way she could, trying to gain control. She would tell him things I had done, turning an event into something horrible like my climbing to the top of our neighbor's forty-foot fig tree, or dancing wildly to the organ in church. She would hound him, telling him if he didn't punish me, I would end up lawless and outrageous, just like Aunt Grace. She would threaten to leave him if he didn't spank me. She would stand behind the door while Dad would sit with me on the bed trying to gather the courage to do what she demanded. Sometimes we would pretend. He would hit the bed, and I would yell my head off.

When Dad finally found a woman he wouldn't stop seeing, Mother left him for good. The day she served him with divorce papers I was eleven years old. We stayed with friends, out of town, so he could collect his things. The next morning when we returned, Dad had taken everything but our clothes and my bed. Our house was completely bare, as though no one lived there.

I was fascinated at seeing the intricacies of Joan's upbringing. In this way I began to look into my own psyche, at things I had never really thought about. After a session with Joan I would drive slowly home or stop by the beach and walk along the sand, contemplating and crying about my past, which was still so vague.

After an emotional outpouring, Joan would be embarrassed. I tried to convince her that nothing she could do would lower the esteem I had for her, but she didn't believe me. She was ashamed. She wanted to leave the past behind. There was nothing I could do. She had lost the weight she wanted. She was fitting into size eight jeans. She looked terrific, and she stopped calling. I missed her terribly.

Whenever a client stopped seeing me, I felt deserted. I knew things about my clients that their husbands or close friends didn't know, but we weren't friends. We were something else. I was taken into the very center of a person's world but allowed only to look. After a massage was over, I returned to my world, void of the intimacy of that family. Being a masseuse has helped me learn a painful lesson; that of detachment.

Smoke

It used to be I couldn't be with anyone, man or woman (especially people I admired,) without becoming self-conscious and shy. Being a masseuse got me past that. I had a role, a purpose and a routine that gave me a way of being. My contact with a client lasted only an hour. If I told a story and didn't get a response like "Um hum" or "That reminds me of" or "Ain't it the truth," I knew to be quiet.

I found I had a lot to say when massaging. When I began to massage a client, the walls of fear and insecurity that kept me isolated began to melt. It was easy for me to talk with the heat of my clients coming up through my hands. The warmth of their bodies reassured me. Even now when I have something difficult to say to someone, I ask, "Can I massage you?"

Sometimes my clients and I just listened to music. Sometimes we were silent, and I would massage them until they asked a question like "Well, what happened next?" Or "Did she find out who he was having an affair with?" Or "What do you think about facelifts?" Or "Did they ever find that boat?" There were stories I told and stories I didn't, and many got mixed up because there were parts of them I could tell and parts I didn't want anyone to know about.

One of the stories I told my clients when they wanted to

quit some bad habit was how I quit smoking. I began by telling them what a glorious habit this was for me and how I discovered cigarettes when I was sixteen. I loved cigarettes from the first moment I took smoke into my lungs. I loved the sensation of lighting the cigarette, blowing out the match, sucking, inhaling, holding my breath, exhaling and tapping the ash. I loved the way cigarettes gave me something to do when I hadn't the faintest idea of what to do. I loved how smoking gave me something wordless to do with people. I loved the way it took my white, blue-eyed, blond-hair, Grace Kelly look and gave the world the impression that I had a black leather jacket waiting for me in my closet.

I loved the way smoking seemed to burn a hole in my ozone and gain attention from all those dangerous boys on motorcycles who had ignored me before. I loved the way my mother hated my smoking, the panic that would rise in her voice when she caught me. Through my smoking I could say, "Fuck you, Mom," in a brand new language.

I was addicted to smoking in no time. Countless times over the next twelve years I tried to quit smoking. During my last year in high school, I tried to quit when I played Anna in *The King and I* so that I could sing better. I tried to quit smoking for my husband, who said he wouldn't marry me unless I did. But I smoked in my wedding dress after walking down the aisle and in the bathroom at the reception hall. I tried to quit when I was pregnant with Russell, and I tried to quit each day of the whole eight months I nursed him, just "not today, not today." I tried to quit smoking when I became a hippie and ran a health food restaurant in Boston. But I was still smoking when I went into the convent in 1975.

I had not been behind the convent walls for ten minutes before I had a nicotine attack. I wanted a cigarette so badly I thought I would die. I didn't want it as badly then as I wanted it later that night, though. I lay in bed after an hour of silence, dinner, two hours of classes, an hour of meditation on my knees–in which God was a deep drag I couldn't have, the Marlboro man I could never reach–and evening prayers.

Everyone in the convent went to bed at eleven. I listened for all the sounds in the house to stop. When all was quiet, I got out of bed and crept toward my cigarettes, which were hidden in a special pocket of my black purse. I heard the breathing of the other women in the room while I pulled a cigarette from my purse. Then I found some matches used to light candles and snuck down two flights of stairs to the basement. There I knew I could light my cigarette without being detected. When I pulled the smoke into my lungs I fell to the floor. I simply collapsed from the rush of nicotine through my veins. Even though I was down and could barely move, I still got the cigarette to my mouth. Yes, I was an addict. I snuck cigarettes in the convent for nine months. Almost my entire stay.

At one point while in the convent, I had to go to the dentist. There in the waiting room was a book called *How To Find Your Soulmate*. I read it in fifteen minutes, though it was clearly not on the prescribed reading list for nuns. The author described in detail how to attract the mate of one's dreams by daily repeating a statement describing the type of person one wanted. He said this daily repetition would draw one's mate through magnetism. He suggested one be as specific as possible, choosing a time and date for this meeting.

I decided to test this theory out on the impossible task of quitting smoking. That very day I created my cigarette affirmation: "On November 8th at midnight I will no longer smoke cigarettes, for the raising of my consciousness and the health of my entire being in accordance with God's divine law, and with the help of all the saints and sages of all ages, Amen." I wasn't taking any chances of offending anyone. November 8th was three months away, which made me feel comfortable. Maybe it would never arrive. Maybe I would die and I wouldn't have to quit.

But of course I didn't die and the day came. On the evening of November 8th, 1975, I smoked my brains out until midnight. The next day I awoke and had no desire for a cigarette at all. In fact, quitting was a relief. Finally, I could stop repeating the affirmation.

When I became a masseuse, I told this story to everyone who wanted to change anything. But I left out a lot. I left out the fact that I was a nun when I quit. I left out the fact that I had been in a foster home when I started smoking—a home where my mother sent me when I became a wild person she could not control. I just gave my clients the bare bones of the story.

Years later I wrote an article about using affirmations; the process of repeating what one wants over and over. The article was called "Affirmation Backfire." In it I described how everything I got through repeating affirmations brought with it things I didn't want, too. For instance, instead of smoking I now had a coffee addiction I couldn't get rid of, and when I later did affirmations for a soulmate and asked for someone who was tall, he came along, but he was six foot five, which was ridiculous for my height. He was also so beautiful I was jealous the entire time I was with him, and he smoked so much dope I thought he was an addict. Because of my three months of saying the affirmation for this man, I became so bonded it was impossible to get away from him. I was tied to him for eight miserable years.

I told the whole story about doing my "soulmate affirmation" in the article, the point of which was to enjoy and love whatever situation you found yourself in, but no matter. All the mail I received about the article wanted to know, "How do I find a soulmate?"

I learned a lot about human nature through telling my clients this story. Everyone I told this story to, and to my knowledge those who read my article, didn't have the discipline or didn't want the same thing for three months. I guess one has to be desperate. I believe it is the state of longing that creates the "burning bush," whether it be a mate, a home, or quitting a habit that is killing you. I have been surprised to find that most people are terrified to want anything, to make a mistake, or even to get what they want.

Mrs. Adams

While working in the beauty parlor I met Mrs. Adams. She was over eighty and didn't see well. She could not turn her neck to back into a parking place without bumping into everything: the curb, the car behind her, the car in front of her. I drove to her home once a week, carrying my massage table up the stairs to her second story condo where she lived with her husband and a black cat named Farley.

Getting up the stairs was the first of many obstacles to overcome before my massage with Mrs. Adams could begin. Getting someone to open the door was the second. Mrs. Adams was as slow as winter in Maine, and Mr. Adams was deaf. I'd bang away on the door with both fists. The doorbell never worked.

Once inside with my table set up, Mrs. Adams would come in and I'd help her take her girdle off. We had a routine. She would face me and put her hands on my shoulders while I would pull the girdle off her hips and down her legs. Then I'd hold her hand while she stepped out of it, where it lay on the floor like a flat tire.

To get her on the table, I had to boost her up, where she would collapse in a heap. Next I would straighten her out, pulling on her scrawny arms and legs in which muscle tone was only a memory. Then I would lift her shoulders and hips so her flesh could rearrange itself and settle in comfort.

I often wondered how skin as thin as hers could hold the blood. It did an imperfect job at best. Purple marks told of her bumping into walls and doors, perhaps her desk.

Mrs. Adams liked to keep the shades open to let in the sunlight. Once I mentioned that people across the way might see her. This old, proper, southern woman, who had every pillow case and sheet in the linen closet wrapped in its own tissue paper, said, "If they see something they've never seen before, they can throw a shoe at it." This bit of openness was the only discrepancy I ever noticed in this otherwise private, secretive woman.

I think her own body was a secret to her. Once while I was hoisting her up on the table, she caught a glimpse of herself in the mirror, naked. Her breasts fell down to her stomach, saggy bags of empty skin. Her stomach and every inch of her was wrinkled and hanging off the bone. She stopped and stared at herself. I was anxious and felt the need to protect her from the sight.

"Is that me?" she asked.

I felt too awkward to answer. Suddenly, she burst out laughing.

"What a mess," she said with a sigh.

There was a feeling of waiting for death in that house, of things decaying and of secrets, but no one ever mentioned them. Mrs. Adams never told "Papa" the constant pain she was in. "It would have frightened him," she said.

She told me she knew Papa was losing his mind, but she never said a word to him. Instead she set out his checks each week for him to sign because he would not have remembered, and she made secret phone calls to their broker in order to take charge of their money.

Mr. Adams, who had run a business all his life, knew

government secrets, and once had hundreds of people working for him, did not know his wife read the *Wall Street Journal* on the toilet each day so she could look after their investments.

The Mr. and Mrs. Adams had outlived most of their friends and all but one 250-pound son who had moved away. One look at him told me what growing up in this land of secrets had been like. He had buried himself alive with food to keep away the horrible feelings of emotional starvation.

The cloud of secrets affected me, too. When I walked through those doors, I left everything about me that was sexual, alive, intelligent, moist, deep, feline, healing or original at the door. In Mrs. Adams' presence I couldn't remember that I'd read *The Tibetan Book Of The Dead*, Carl Jung, or Fyodor Dostoyevsky. With Mrs. Adams I never mentioned anything more spiritual than *Thoughts for the Day*, by Hazelton. She would talk about the goodness of Jesus, and I would say, "Amen."

But I learned a great deal from Mrs. Adams, and I enjoyed our "visits," as she called them. I learned a different kind of closeness through her, which had nothing to do with agreeing with each other. It was a closeness born of words not said. There was a part of me that understood her sighs, a look, a moment's hesitation, all of which made me love her. I knew for certain I was the only one who saw that look of despair or heard that sigh, and I would have lugged my table up five flights of stairs to give her an hour's comfort. I knew her at her dying time, when all the wisdom she had gained from her life's experiences had been melted down to its essence, when all her endurance, discipline and courage shone in her eyes.

She had the will of a road warrior. She told me that she had allowed herself one cigarette along with one cocktail every day with her husband at 4:00 P.M. for fifty-five years.

Her maid used to say, "There they are, courting again."

I asked, "Didn't you ever want another cigarette or another drink?"

"Oh, yes," she said, "sometimes I wanted one very badly."

Mrs. Adams told me things she had never told another soul. She told me things like, "I never got to ride a bicycle. My parents were poor. Our maid drinks." I wish I'd heard her say just once, "I'm dying of cancer and I'm scared."

After four years of massaging her weekly, she had to stop. It was just too painful for her to get up on the table. I went to see Mrs. Adams after Papa had died. She had called to tell me of his passing. I walked into her exquisite living room where I saw the wallpaper falling down in sheets. Her cataracts had made it impossible for her to see.

There was a chair Mrs. Adams had in her sitting room that I had always loved. Farley had clawed the arms to shreds. The scene made me want to steal everything. At that moment I understood a scene from the movie, *Zorba the Greek*. When the foreign woman died, all the women in the village descended on her house and stripped it to the bone. The women were like furies. They took down her curtains and fought over her shawls, opened the closets and took her slips and stockings, her umbrella, her parrot and her dresser. What they couldn't carry down the stairs they threw over the balcony. Standing in Mrs. Adam's living room, I understood that scene; it was a shout declaring, "I am alive. I still have needs. I am alive. I want."

I never saw Mrs. Adams again. When I went back six months later, no one by that name lived there. I tried to find out what had become of her. I talked to the man who opened the gate for visitors and the manager, but no one knew anything. I felt terrible that I had not come back to visit her sooner. How had so much time slipped away? But then it took time to know how much she meant to me, how her words, her gestures, her stories, had crept into my psyche and taken up residence in my body.

One very late, cold November night, on my way home from massaging a businessman, speeding down Sunset in my beat-up old Honda, massage table in the hatchback, I suddenly knew she was dead. I became aware in a flash of all she had given me; a model of strength and courage to endure and contain all the pain that life brought–to the very end.

There in the car I cried for her and I cried for myself, and I cried for all the women whose upbringing could not let them say, "I have cancer and I'm scared, Papa. Hold me in your arms, and help me bear it."

When I am feeling very old and tired, Mrs. Adams enters my mind and stares at me with her watery, gray eyes, so filled with compassion and pain.

Hoover

I was pounding Pam's cellulite in the massage room of an exercise studio, another place I worked for a while. Instructors shouting, "Kick right, kick left, tighten thighs, breathe," came through the walls. Such frantic energy, such loud noise and so much effort expended to use up the excesses of affluence: too much money, food, time, cars; not enough sex, love or dignity. My back hurt. I wanted to fart. I was feeling sorry for myself.

Just that morning I had been to the doctor. I had had a pelvic inflammatory infection years before, and I thought it had come back. While examining me the doctor said I would never get pregnant again. He told me no more eggs would make it through my mangled fallopian tubes. All those potential children who could show me a different part of myself would die without a breath, without a cry or sound, without limbs or insight or wisdom or anger. I could stop making marks on the calendar.

I was bemoaning my fate when Pam said, "What you need, honey, is a puppy."

When I finished the massage, we headed straight for the pound. We looked in cages at dogs of all sizes. Then a volunteer walked by with an eight-week-old, furry, fuzzy, cuddly black puppy with four white feet and a white chest. He fit in both hands and was the only one in a litter of eight who hadn't found a home. He had arrived at the pound that afternoon. He was a Sagittarius, but at the moment he did not look much like a philosopher. He was bewildered and had funny ears that flopped when he walked. I called him Mr. Floppy Ears until his real name came to me, Hoover. He was not named after the dam or the president but after the vacuum cleaner because he ate everything on the floor.

The first morning of Hoover's life with me a woman came to my house for massage. I put Hoover in the kitchen and closed the door. He began to whine and yap, pressing his black nose under the door, trying to get a whiff of me. Well, of course he got in and took up residence under my table, where he lay during every massage I did at home for the next thirteen years.

Hoover was the main reason I worked at home. I remember when I took a full time job in a facial salon. Every day while I was at work, Hoover took everything he could get his teeth on in the garage, carted it out to the driveway and tore it to shreds. After three days of this I got the message, quit my job and came home where I belonged.

When I massaged someone in their home, Hoover came along and sat on my client's shaded front porch or under a tree with a bowl full of water and waited for me. Unleashed, he would wait for hours. He would not tolerate being tied up unless he was in a public place where he might be mistaken for a stray. And there were lots of places I chose not to go because I couldn't take Hoover with me.

He was quiet as a mouse under my table. He knew scratching and licking were not allowed. If he had to do such things, he left the room. Sometimes when I was massaging, he put his head on my foot, a gesture of friendship and devotion.

Hoover was a great teacher of mine. Even when he was very old, blind and deaf, with arthritis in his back, he communicated with me through signs, the language of dogs. When I heard him pant, I knew his water bowl was dry. There was a certain look that meant, "Take me for a walk," and another that said, "You forgot my dog bone." There were times he let me know he needed to run wild or rummage through trash to remember he was really a wolf, and times he needed to be on a leash to keep him safe from being picked up and taken back to the pound, or to keep him with me in large crowds. He taught me the language of grunts, tail wags, side steps and squeals. He had a way of licking his lips to ask for a bite.

Sometimes I would get angry at Hoover, usually when he didn't do what I wanted fast enough. It was usually because I hadn't taken the time to be clear or make sure he was listening. Anger with him was always a mistake. When he knew what I wanted, he always tried to comply out of his love for me, not fear. Fear of me made Hoover a coward. It made him sneaky and resentful. At these times, when I'd been bad, I had to sit with him a long time, scratching his ears and his favorite place on his back, or rubbing his tummy, or just sitting with him quietly. He was always forgiving. If I carried on too long, he got up and left in disgust.

After fifteen years of being my familiar, he died on a camping trip with me in the Rocky Mountains of Colorado. When my friends found out about his death, I received fifty condolence letters. He died in my arms, his furry face next to mine.

I took his ashes with me back to California, and my mother, Russell and I did a ceremony for him. We went into the woods, sat around his box of ashes, and told "Hoover stories." Then we scattered his ashes, crying our eyes out.

Because I did not get to raise my son, there was an

emptiness, an incompleteness in me that had to be filled by a similar task. Taking care of Hoover completed that task. It was difficult keeping a dog in L.A. Some places I lived didn't take dogs. I had to hide him when the manager came by. I had to walk him four times a day so he could get his exercise and be comfortable. The last year-and-a-half of his life he was sick, needing medication and special care. I performed these acts as though he were my child. One man wrote to me, "Watching you and Hoover was like watching one of the great loves of the world; like Hepburn and Tracy, Anthony and Cleo."

Through Hoover I proved to myself that I could go to the end with something. I proved to myself that I could endure and that someone could not only survive, but thrive under my care. I proved to myself that I could be faithful, that I could love.

Sam and Babe

The only legitimate call I ever got from my ad in the phone book was from Sam and Babe. Sam was a retired, ex-rock star, and Babe was an ex-hair dresser and now Sam's gorgeous wife.

They bought a house while I knew them. It was the first house Babe had ever owned and I watched her turn it into paradise itself. She could turn a broken mirror into a beautiful, intricately patterned decoration. I watched her build a seventy-five-foot by seven-foot stone wall along one side of their property, all decorated with the ends of wine bottles. She planted cactus gardens that looked as if she had been to design school. She had amazing vision. Her walkways sprouted arbors, sweet peas crawling up and over them, laughing and waving red, lavender, purple and white blossoms. She could plan a stairway in her head, create a wall of mirrors and design a patio with a fountain before I'd finished massaging her feet. Once, after a massage, she gave me a jar of homemade moisture cream she'd made from avocados and cucumbers she'd grown from seeds.

Sam would watch her creative antics and grin. When he got bored, he would hire a plane to take them to Vegas or Europe or Bali. They would come back having taken rolls of film. She had the pictures developed and put them in albums edged with pictures from last year's calendars, sending copies to farmers and natives she'd met along the way.

For Sam's thirty-fifth birthday he and Babe invited me, a friend, and six other couples to Maui, all expenses paid, to help him celebrate. It was to be a party to end all parties. Every drug known to the Western world made it past those dogs at the airport.

I had never used cocaine before. My life was already full of things to resist. I knew I had to resist seminars and gurus with long beards and hypnotic eyes. Gurus had convinced me to give away everything I owned–twice–and I was onto them. I'd learned to resist lunch for laughs with old lovers. I'd learned to resist saying, "I'm leaving," to make a point, and coffee before bed time. I'd learned to resist the open road, and the idea that it would all be better in Oregon, and I now resisted double-dutch-frosted chocolate cake on a sunny afternoon when I thought nothing could depress me.

In the past when I'd been offered cocaine, I had said, "No, thank you." But in Hawaii, on a warm, balmy night, in a dining room we had to ourselves, right on the water, with twelve other people doing it, and waiters tooting lines right off plates on the table, I said, "Oh, what the hell."

I had a line of coke and became addicted.

When we got back to California, Sam and Babe gave me cocaine as a tip for massage. The cocaine sang to me like a siren until it was gone. Anytime Sam and Babe called, I'd go, and from time to time I'd just show up. I would try to tie myself to the mast of sanity with thoughts of caution and fear, incriminations. Then I remembered what Henry Miller said: "He didn't believe in moderation nor good sense nor anything that was inhibitory. He believed in going the whole hog and then taking his punishment." Perhaps Henry Miller had a sense of Self that allowed him to have confidence and dignity even with his head in the toilet.

I'd go out on a limb with my head full of "Millerisms" or some borrowed ideology that said, "Live life to the fullest," then come crashing down with no inner confidence to support such thoughts. I could go "the whole hog," but I couldn't take the punishment the drugs dished out. The highs would lead to the lows, and the lows would make me crave the highs again, and eventually the cycle would lead me to despair and anxiety attacks. When I took drugs or drank too much, I didn't know what I thought or felt or whether it was me or the drugs talking. I needed a middle path, a harbor, a safe port, shallow warm water that was quiet and calm so that I could begin to hear the faint sound of my Self again.

I got to a point where I couldn't tell which voice in my head was telling me the truth. The only thing that kept me going was massaging my clients. Mrs. Adams kept me alive, along with Mrs. Crow, Mrs. Cook, Mrs. Douglas, Mrs. Vanderbuilt, Mrs. Earl, the general's wife, the soap opera writer and all the others. As soon as I lay my hands on their bodies, whatever was going on with me would fade into the background, and I would be at peace. I saw most of my clients every week. I usually had six appointments a day. I loved them. They loved me. They gave me birthday presents and extra money at Christmas. I never came to them drunk or stoned. They were my shrine, my offering, my prayer. I never wanted to pollute them.

Each morning I awoke and drank a liver flush, which consisted of fresh orange juice, a lemon, garlic and olive oil. After yoga and meditation I went to work. When I came home, after talking all day about the importance of diet, relaxation and self-cleansing, I would smoke a cigarette, drink a glass of wine, and snort a line of cocaine off the mirror under the sofa and hope no one called.

When Russell came to visit for the summer, I stopped

doing everything: smoking, drinking, cocaine. I repeated affirmations to clean up my act, starting a month in advance of Russell's arrival. I lived on the beach so when he came home, and I wasn't with clients, I would watch him ride the waves on his boogie board. His white East Coast body would turn golden brown, and the tension in his face caused by school and his drive to become a star athlete would relax.

During this time, I met a woman who taught re-evaluation counseling. I took her course. One of her exercises was to tell another person in the group our life story. Then the other person would tell their story. Telling my life story in an hour I heard myself say many times, "My life began to fall apart as I was drinking too much."

Soon after this class I took another workshop to become a rebirther. Rebirthing is a technique used to clear up birth trauma. The workshop was held every evening for several weeks in the large living room of one of the participants. One evening we went around the room and each person was given the chance to talk about what the workshop had done for them, and I heard myself say, "It has given me a place to come every night so I wouldn't have to drink."

The words were out before I could catch them. They were like wild birds, or bats let out from a dark cave. Blind, they darted about the room, their wings beating the air hysterically. In the silence, which was filled with fear, remorse, pity, mercy, compassion, benevolence, someone asked, "Are you an alcoholic?" And I said, "Yes, yes I am."

It was more than a year before I could stop drinking and using drugs. Just admitting my addiction to sixty people wasn't enough. I had to reach complete and utter defeat. It was like crawling down into the bowels of the earth, where I would beg and bargain and plead with the gods of the underworld to let me have just one more night with my beloved. The beloved who would lead me to feelings of union, where I could live out my fantasies in peace. Where I could feel I was a great soul chained, like Prometheus, to the rock of despair. And all was romantic, not tragic.

In the evening after work when the drug or wine wore off, I would go to the bathroom and wash my hands and mouth, put on cold cream and get into bed, praying for sleep. Sometimes I got up five or six times a night and went through my routine: up and down the driveway, drinking, washing, cleansing, doing cocaine, feeling calm or anxiety, finding the matches, one more cigarette left.

I remembered my Aunt Grace, who came to live with Mom and me every time she got a divorce. I remembered hearing my aunt, seeing light under her door, the smell of her cigarette coming through the heating vent, the quiet foot steps down the hall to the kitchen, the squeak of the cupboard door where the liquor was kept. Sometimes I would hear her cry. She died at forty-five of liver disease. Had her ghost entered me? Was I walking her hall, performing her ritual, reaching the end for both of us? Perhaps I wasn't trying to escape my life. Maybe I was being a loyal servant to the dead.

Eventually I reached the bottom of the pit. It happened one night when I was at a party with all my healing friends. Everyone was outside in the hot tub under a starry sky. Candles accented the landscaped patio, and little lanterns gave light to a path lined with huge oak trees. I walked through the kitchen and living room drinking every half-filled glass of beer, wine, whiskey, and vodka–even the ones with cigarettes put out in them. When I was through and there was not a drop of liquor left in the house, I realized I was stone sober.

It reminded me of a time I took a hit of acid and had nothing happen. Then taking another hit. A minute later the first hit came on so strong it took my head off. I realized then that the second tab hadn't even hit my stomach, and then there was a moment of clarity. Right before I went off to other worlds, when I knew I had taken on much more than my nervous system had ever dealt with, and there was no way out but through the asshole of the universe, there was only time to think, "OOPS."

I awoke in my bed at three in the morning wondering how I got there. I could not move, or the creatures right before my eyes would have pounced on me. They were circling me with their rat eyes, making me hold still. I could hear them at my throat and deep inside my guts, sucking, gnawing, hissing.

I had hallucinations for three days before I was able to call my friend Annette, who told me about a group of people who could help me. I was ready. I would do anything not to repeat that experience of anguish, fear and despair. And from that night in 1980 I stopped all use of drugs or alcohol, and, by the grace of God, have not gone back to them.

And I let go of my clients Sam and Babe. Just hearing their voices on the phone made me want cocaine, and I was afraid to see them, afraid that in their presence I would lose my resolve to stop. I came to the realization that in order to let go of the worst of life, sometimes one has to let go of the best that was also a part of the bad.

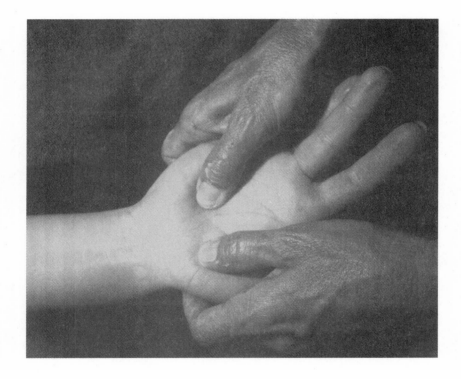

Massage of The Heart

Swedish massage is "massage of the heart." To give one is an act of love. It is caring for that person as if they were your own beloved. At times my clients' little fingers are as dear to me as my son's. Their aching backs become my mother's; their feet, my guru's. When I've reached that point of love within myself, I feel I have connected with the God in us both. It is an act of communion and takes sinking into the silence to get there. It is rare to find someone who can be that still. During massage, when getting close to deep silence, people will frequently leap up or say something desperate or simply fall asleep. But sometimes it happens that both of us get to that place of indescribable stillness, and we are able to feel the magic of connection and become one.

In the past, after a week of massaging, I often couldn't flex my hands, and sometimes I couldn't talk. My mind was a jumble; my feelings a mish-mash of all the people I'd worked on. Someone might come to me with a headache, and I could feel it come up my arms as I massaged her neck. She would leave feeling great, but I would have her migraine for three days.

It took me years to learn how not to take my client's problem. Before I learned, my knuckles would ache from arthritis, and my back would go out on a regular basis. A chiropractor showed me how to "block" my hips, which temporarily relieved my back. This technique consisted of lying on my stomach and putting a block of wood under each hip, which allowed my pelvis and lower back to sink into the proper position. To accomplish this on my own, I carried a pair of high-heeled shoes in my sheet-and-oil bag so I could block myself after each massage.

After working all day (sometimes until late at night), I would wash and hang my massage sheets. If I didn't wash them right away, or if I put them in a dryer, the almond oil I used for massage would turn rancid and the sheets would smell. So I hung them on a clothesline in the backyard. Feeling my way in the dark, a bag of clothespins around my neck, I hung the sheets so straight that they dried without a wrinkle. Sometimes I sat on the stairs beneath the sheets thinking of them collecting the essence of the night, the light of the moon and stars, the dew of the morning, and the sun, which dried them into white flags.

When I was exhausted I would go to a stream near Ojai and put up a tent. I listened to the water and watched the way the light came through the leaves of the large oaks. I had never been so happy. I had never been so busy or felt so needed. I had never dreamed that I would be so useful to so many people or so loved. I had never dreamed there could be a life in which I fit so well.

I had been working since high school at all sorts of things. I had so many jobs I couldn't count them all. In high school I had a job as a file clerk during the summer at a title company; it took them months to find anything after I left. I had worked in the complaint department of a condominium management company. People would call when the gardeners mowed up their welcome mats, or they found a frog in their toilet, or their roof leaked. I had been a waitress in a coffee shop on the early morning shift. Once I worked as a nurse's aid, first working in a retirement home and then working in a delivery room. I had been a devoted disciple of an Indian guru, and a hippie in a commune that started the first health food restaurant in Boston.

I never wanted to be a masseuse. I never wanted to be a file clerk either, or a waitress or a secretary. I never wanted to be a nurse's aide wiping old people's bottoms and bathing what was left of their bodies, watching them as they slowly went insane from medication and neglect. I never wanted to be a nun either–I loved sex too much; I loved life too much. I never wanted to be a wife to a workaholic or a mother who abandoned her son or alone without money, drunk. These are not things young girls dream of being.

I'd write home to Mom, "Be patient. God isn't through with me yet." Or, "I'm collecting stories to tell my grandchildren."

My parents couldn't understand where my restlessness came from. They had moved very little. My father had the same job for forty years. Perhaps I inherited the restless genes he wouldn't use.

I believe children inherit the unresolved lives of their ancestors. On my father's side I believe I inherited his lust for pleasure and freedom. As for my mother, I believe I inherited her loneliness and her broken heart, her terror of life, and her certainty of abandonment. My mother's mother died when she was four. She was raised by resentful aunts and uncles who didn't want her and then by strangers in foster homes and orphanages. My mother had never been mothered. Now at eighty she has finally recovered and is enjoying her life. When she was sixty-eight her husband died and she sold the farm and moved to Guadalajara. She spends six months out of the year there, driving there herself. She was once held up and robbed by banditos. She mentioned it to me on page three of her weekly letter. Nothing stops her.

It took years to develop my theory of inheritance. It took one heartbreaking, out-on-a-limb adventure after another. Like the time I found myself standing in the middle of a jungle in Puerto Rico. I was looking for a man named Manuel whom I had never met. I was in a torrential rain that made noon black as night. As I stood there on a steep path turned into a stream by the rain, there was a clap of thunder that shook the earth and a flash of lightning in which I could see the pitiful predicament I was in. I had a moment of clarity. All of a sudden I realized, "I'm not paddling with both oars. I've been rendered an idiot by conflicting needs and desires. I am racing in two opposite directions as fast as I can, which keeps me in the same place."

But by the time I'd have a thought like that, it would be too late to stop the inevitable catastrophe and I'd have to start life all over again with a new man, in a new town, looking for a new house and a new job. It's hard, even now, to stop looking for "For Rent" signs. It's become one of my "built-ins."

And for some reason, after finding a new man, a new house and a new job; after the pieces of my life began to settle, I would be off again. I could not bear the anxiety that came with safety. The intensity of pain, confusion and chaos was home to me, and try as I might, I could not hold still. I moved at least fifty times in ten years. I lived in San Jose, Boston, Arlington, Cambridge, Del Mar, Roxbury, Portland and St. Petersburg. I also moved to San Francisco, Sausalito, Kennedy Meadows, Santa Barbara and Los Angeles, and many places in between.

Then by accident I became a masseuse. My clients and my profession became my refuge, a safe harbor where I could anchor my restless self. As a masseuse I could live in one town and still keep traveling from one door to the next, collecting stories, comforting clients and laughing with them. I was astonished at my good fortune in having found a job in which I got paid to touch someone. To find that I was good at something, that I had a profession I loved, a profession that supported me in a manner I had never dreamed of achieving on my own, held as much wonder for me as a miracle.

As a masseuse the separateness between body and religion began to diminish, and those two warring factions of my psyche began to merge until I saw the body as a temple, a place to worship, a holy place that houses Spirit and Soul. And I began to have confidence for the first time in my life.

And my son was proud of me.

Stories

Every client has a story that lives in my memory. I have blue stories, scarlet and magenta stories, yellow stories, black stories, chartreuse and deep cobalt stories. There are stories as pale as see-through lime green and as red as blood.

I heard about affairs and sexual soirees I couldn't believe. One woman told me she had been making love in a row boat in the middle of a lake when the boat tipped over, but she and her lover couldn't stop. When they finally climaxed and looked for the boat it was nowhere to be found. They never did find it. They had to swim to shore.

Riley is another story. He was a real estate man and his entire office got together and sent him to me for a massage. When he was on his back, I put one hand on his forehead and the other hand under his neck. His pulse pounded with caffeine. "How much coffee do you drink a day?" I asked.

He was amazed I knew he drank coffee. Already he thought I was a wizard. "Maybe thirty cups a day," he said.

I asked him to keep track for several days and then cut the amount in half. He came back the following week and said he was down to twenty cups. In a few months he was down to five cups. Then he gave it up all together. With this feat behind him, he discovered he hated selling real estate, that it was only his consumption of large quantities of caffeine that had kept him tied to his desk and the phone for so long.

After six months Riley came to me for his last massage and told me he was moving his family to Bali where he was starting a scuba diving business. He said it was a lifelong dream that he had finally slowed down enough to remember.

Katherine

Over the years many of my clients gave me clothes. They gave me boxes full of nighties, sweaters, skirts, boots, blouses, T-shirts and swimming suits. I ended up with the most unusual wardrobe. My clients all had different styles, most of which I would never have chosen, or could not afford, and none of which went together. So what I couldn't use I passed along to friends. I opened each package with awe, as if it were buried treasure, suddenly aware of all I'd never had.

Katherine came to me for cellulite massage while I was working at the beauty shop. She was a wealthy woman who had been a debutante. When she went shopping and found something she liked, she bought it in all colors. As a result she had tons of clothes, which flowed out of her closets and drawers like lava. She gave me thousands of dollars' worth of clothes.

Katherine came to me twice a week for a year. By the time every ounce of cellulite had vanished from her thighs and buttocks, we had become very close. It was unthinkable that we should never see each other again, so we became walking buddies.

Katherine was the first client I became friends with. I never had a friend so unlike me. At first we had nothing in common except the desire to be thin and our love of walking by the ocean. As time went by, we walked every day, and every day we found more to share. As we talked, we walked faster and faster until we were doing four miles in an hour, barefoot.

We got to know each other in a whole new way. It is one thing to know someone professionally and quite another to know them privately. At first we talked about recipes for Thanksgiving and Christmas. She told me her traditions and gave me the recipe for her family's favorite pie, chocolate mousse pie, which became my son's favorite when I made it for him. In fact, I gave sugar rushes to much of the health community with Katherine's chocolate mousse pie.

I was intimidated by Katherine at first. I envied her four children and her house in the mountains. I envied her clothes, her experience as a debutante, and her wealthy beginnings. I envied her assurance in the world and the fact that she did not have to work. I loved her long fingernails, which were always polished and without chips. And I loved her hair, which always looked neat, and I loved that she drove out of town to get it trimmed. I envied her because she could wear false eyelashes without a hint of self-consciousness, and she always had a new jogging suit. I envied her Porsche, which was always clean and full of gas.

Katherine was the most feminine woman I'd ever met and cheerful as a canary. I was thirty-four and she was forty. She became the older, more mature woman in my life. I was ashamed of my life when I compared it with hers. In my eyes God loved her more than me. My mother was right; I had wasted my life.

We'd been walking for over a year and it was Christmas again. We met one morning as usual, but Katherine didn't smile when she got out of her car. She'd been crying. She was a nervous wreck. Her husband was going away on business, and he hadn't asked her to go with him. She hated him for this, and besides he wasn't leaving her enough money. What should she do? On top of that, her youngest son was in trouble. He had stolen a car, and he wouldn't go to school. He wouldn't speak to anyone in the family.

The next morning she was worse. While walking, she broke down and cried. She felt that she was losing her mind. Her family was driving her crazy. I don't believe we had ever had a silent moment between us until I realized, in the throes of her distress, that her husband was having an affair with one of my clients.

I remember my client talking about a man she loved–a man who was married and had a family. She said she was going with him to Europe over the holidays and felt guilty, but not guilty enough not to go. In one moment, as we walked in the rainbow colors the sunset showered on the tide, it all fit together.

It was hard for me to tell Katherine. I tried to open my mouth with just a beginning word, but it was as if my body and my voice had a mind of their own. It was like they were protecting her against my will. When I was finally able to tell her, "Your husband is having an affair," our walk became dark as death. Katherine was like a tree with no roots, like a fish in air. Her philosophy did not work without money or position or a husband to give her these things.

It was a turning point for me, a recognition of myself and my value. For the first time I shared with another person how I survived a life of loss–the philosophy and knowledge I had gained while trying to make it until dawn after the loss of my son, or getting over the separation anxiety after leaving a man. I turned on my light and became Katherine's guide. I taught her to say the mantra, "Divine Mother, have mercy on me." We said it together five-hundred times in a row on our walks, and then she would say it in her kitchen, alone at her table.

I taught her how to listen to her breath and try to breathe with the rhythm of her heartbeat at night when she couldn't stand her thoughts. I suggested she get a notebook and write out her anger and pain as if they were a story, and to take the feelings, which were eating a hole in her belly, to bed with her and to comfort herself like a hurt child. I suggested she take those feeling of loneliness and grief and watch their passage until they settled down to the bottom of her.

And when she was totally opened by despair, we became true sisters. We had both revealed ourselves to each other. There was no fear between us. She knew everything I could remember that I had thought or done, and I knew her.

We tried to blend our lives more, but it never really worked. Katherine didn't fit into my life, and I didn't fit into hers. Her house made me uncomfortable; she never liked "long-hairs." I didn't like the relationship she had with her children or the way she treated her dogs. It was only when we walked along the beach, waves crashing, stepping through water or along the hard sand, around the points and over the rocks, always barefoot, that we were in perfect harmony.

After her divorce, Katherine moved away. We wrote Christmas cards for several years and telephoned when we needed to hear each other's voice for comfort. I often wonder if she still has need of the prayers I taught her. I wonder if her concept of life deepened from knowing me. She taught me that no matter what our position in life, no matter how much money we have, how fine our house or our clothes, in our hearts we are all the same.

I love you, Katherine, wherever you are. I hope you are well, my sister.

Stewart

Stewart, an accountant, was the husband of a client who had purchased ten massages for him for his birthday. He didn't even want one, but he came to keep peace in his home. He arrived at my front door right on time, wearing a three-piece suit, reeking from the cares and tensions at work.

I told him what to do with his clothes, how to lie down on the table under the towel, and left him to do so in private. When I returned, he was on top of the towel in his boxer shorts, which was just fine with me.

I put one hand under his neck and one on his forehead. I felt his neck and head pulsing with tension. His skin was damp and clammy. His body odor was awful. This man was stressed to the max. I had to get rid of all my squeamishness in order to touch his smelly feet. I told him his only job now was to let go and relax. I might as well have saved my breath. Every time I touched him, he tensed. I tried to talk pleasantly about this and that—his wife, his birthday, what he did for a living. Nothing I said or did made Stewart relax or feel comfortable. He began to fidget and sigh with annoyance. It was as if I were causing him pain.

The following week he began our session with a question, "What do you think is the key to relaxation?" I thought it was a trick question; he merely wanted to find an opening to ridicule all my ideas, but I played the game. I told him, "Being at peace with your past, a good diet, and a good mental outlook."

"Well, that's pretty simplistic," he said.

At that point I was massaging his leg, which he held up in the air. Irritably I said, "Stewart, you are doing everything possible to ensure that absolutely nothing will happen here. I am trying my best to relax you and make this a pleasant experience. If you don't want a massage or don't like me, I will be happy to refund your money."

I was astonished at my outburst. I had never snarled at a client before. Silence and tension filled the room. Finally he said, "I'm sorry. I know I'm tense and irritable. That's why I'm here. I can't get along with anyone. I have promised my wife I will go to one of those weekend seminars to discover why I'm so angry. I'm going to it next weekend. Let's see if I'm better after that. Give me one more chance."

He could have knocked me over with a feather.

The next week I could not believe my eyes when I opened the door to greet him. Stewart's face was filled with a mixture of aliveness, softness and strength. He gave me a hug and apologized for being so uncooperative. He said he had spent the whole weekend crying about his childhood and the resentments he had toward his parents. The wasted time. Over the weekend he had called his parents and told them his insights and asked their forgiveness for making them responsible for his miserable life. He had made amends to his wife and his stepdaughter.

When I put my hands on his body, there was not a bit of tension anywhere. It validated my belief that the body is the physical manifestation of all our thoughts and emotions. Tension is a result of negative thinking, of unexpressed feelings and unexpressed fears we don't face. It seems Stewart's tears and forgiveness had cleaned him out. If I had not felt the difference with my own hands, I would not have believed it. I did not think a person could change himself so drastically in one weekend, but Stewart was proof that change could happen quickly. Now I know that anything is possible. From that moment on, Stewart's life continued to change. I continued to massage him and his wife for many years.

Jan, Muktananda and Kirpal Singh

My favorite client was a starlet named Jan. Jan and I inspired each other to expose our most outrageous stories, and we would howl with laughter. We roared and shrieked over our stories, getting more and more animated with each tale. Sometimes Jan had to relax from the sheer exhaustion of laughing.

Some people are amazed when I tell them about talking or laughing during a massage. They assume that to get the benefit of touching, a person has to be quiet. It has been my experience that people often need to talk before allowing intimacy. So the more intimate I am with my own stories, the safer another person is in sharing their stories. It isn't about the person on the table baring their soul; it is about what we create together. If I have a break through about my life while I am doing a massage, my client can benefit from it too.

I don't believe I ever enjoyed massaging anyone more than Jan. I usually massaged Jan in her living room, but one day she had all her curtains out being cleaned so we decided to use the bedroom. I carried my table into her bedroom and stopped dead still. There on her dresser was a picture of Muktananda, an Indian guru I had great respect for. I said to her, "Oh, you know Muktananda? I used to have an Indian guru."

She asked, "What was his name?"

"Kirpal Singh," I said, and told her then and there the whole story.

I learned about Kirpal Singh after my divorce. I was hopelessly lost and living on the streets in Cambridge, Massachusetts, selling *The Phoenix* and *Boston After Dark* newspapers at Harvard Square. One day a young man who bought a paper hung around to talk. Soon the subject of spirituality came up, and he invited me to his home, where he lived with a number of others who followed the teachings of an East Indian man. I was curious and hungry, so I went that night to have dinner and hear more about this person.

There were pictures of Kirpal Singh all over the place. He wore a turban and had piercing, blue eyes that shot out of his dark face. His beard was long and gray. All in all he was the perfect-looking guru.

At dinner I was told that none of Kirpal Singh's devotees ate meat or eggs and sex was not allowed except to make babies. Devotees had to meditate two hours a day and listen to something called the "sound current" for another hour. Also, no one was allowed to consume a single drop of alcohol or a tab of MDA, LSD, peyote, pot, hash, morning glory seeds, black beauties, reds, downers, arsenal cans, nothing! And with that I walked out the door.

In bed that night as I was dropping off to sleep, I thought, "What was that man's name?" Sometime during the night, as I lay fast asleep, his face suddenly stared into mine and he said, "My name is Kirpal Singh."

I sat up in bed, immediately awake. Kirpal Singh had visited me, of that I was sure. I was sure he had heard me ask his name, and somehow he had found me. I could still feel his presence in the room. I was shaken and impressed and thought, "Well, maybe I'll give his group a try."

I found out where the meditation groups met in Boston and began going several times a week. Except for a few flickering candles, I sat in the dark with thirty or forty other people. I tried to concentrate on the silence and not think of beer. To me everyone who went there seemed holy. They all seemed to have a particular, knowing smile that reeked of gratitude and well-being. I would look in the mirror and practice that smile, trying to fathom its meaning, trying to see how it felt on my face. It never felt natural. Usually I left the meditation hall and headed straight to the liquor store. Perhaps that's why I never got the smile right.

Three months later Kirpal Singh came to Boston. From the first moment I saw him I loved him. I went to hear him lecture. I couldn't understand a word he said because of his Indian accent, but I experienced a peace I had never known. He was going to give initiation to those who had been able to refrain from sex, drugs, alcohol, mayonnaise with eggs, meat, lying, cheating, etc., etc. for three months. I thought, "Maybe I can sneak in and he won't know." I was desperate to stop drinking. I was miserable at the loss of my son. I visited Russell every other weekend, and I thought, "I may not be able to have a normal life my mom could appreciate, but maybe I can rise above it all by becoming enlightened."

At the initiation ceremony Kirpal Singh went around the room touching people. When he came to me, his eyes closed, for what seemed like hours, before he touched me. He had accepted me as a student. I felt while he stood before me that he was reading my "karma"–all past mistakes and good deeds–and was deciding whether to take me on as a project. I felt he knew I hadn't done a thing according to his rules, but for some reason he didn't refuse me.

For months I said the mantra. I got up at five in the morning wherever I was: with a lover, in a car, on a train, hitchhiking to Connecticut to visit my son. I'd get up at dawn and say that mantra for two hours all wrapped up in "Blue Cloud." I went to meditations and services in Boston where his tapes where played, and I did my best to concentrate. I kept a record of all my transgressions, which, according to the rules, I was to send to Kirpal Singh every three months.

After three months, however, I was ready to give up on the whole project. Along with my daily diaries I sent a note to Kirpal Singh telling him I couldn't possibly follow all his rules and asked him to please cross me off his list. He wrote back in his shaky, eighty-two-year old hand, "Once a student is connected to the guru, you cannot be disconnected. Do not worry. Just do the best you can."

I tried to live life his way a bit longer and keep the diaries, but eventually I gave up. I later heard that Kirpal Singh had died in 1972.

Before becoming a masseuse in 1975, I went to a psychic in Los Angeles. The psychic asked, "Who is this Indian man hovering over you? He's like a mother hen to you. He loves you very much." I told her, "I can't imagine who it could be."

Since I had met Kirpal Singh, I had been in the convent and on all those drugs, which had scrambled my memory. I had forgotten all about him. I had followed many gurus. I knew the words to all of Yogananda's songs, the Indian man who had started the Self Realization Fellowship. I had been to lectures by Muktananda. I had studied with a group involved with the teachings of Gurdjieff and Ospensky. I had done mantra to Mother Mary. To each saint I prayed, "Give me back my son. Help me stop drinking. Take away this pain that lives in my heart."

Later on another psychic told me there was an Indian man whose constant prayers had saved my life. I was grateful for his help, whoever he was, but I still couldn't imagine who he might be.

Suddenly, on the day I massaged Jan in her bedroom,

looking at Muktananda's picture, I knew the psychics had seen Kirpal Singh. He hadn't left me. Jan and I stared at each other. Then we both screamed and hugged and jumped up and down.

I went home from Jan's house high as a kite. I took a bath and tried to remember my mantra. It had been nine years since I'd said it. I didn't know a soul who knew about Kirpal Singh or how to contact one of his groups. I prayed to Muktananda and then to Yogananda to help me.

A friend suggested, "Why don't you go to the Bodhi Tree book store and see if someone there knows about him?" That same day I found myself on Melrose Avenue in front of the Bodhi Tree, and I walked in the door. I asked the clerk if they had a book by Kirpal Singh, and she said, "Of course."

"Well, I really don't want his book. I want to know who took over his clientele," I said.

A man standing nearby asked, "Why are you asking about Kirpal Singh?" I replied, "Because I was initiated by him many years ago. I've forgotten everything he taught me, and now I want to remember."

"Well," he said, "what do you want to know? I lived with him for six months before he died."

Hummm.

On the wall next to us was Kirpal Singh's picture, the same one I'd seen in the house in Cambridge. Robert, the man in the bookstore, and I went outside, and on the lawn in front of the Bodhi Tree he whispered the mantra over and over until I had it by heart. The next day Robert came to my house with books and tapes and addresses of meditation groups throughout L. A., which I went to for eight months.

But all that meditating made me crazy, the diet made me hypoglycemic, and in the end I couldn't stand the idea of another man telling me what to do with my life. What I couldn't stand the most was that little smile on the faces of all the devotees. I needed my own path. I needed to suffer to finally turn toward the light, not in despair, but in confidence that I could endure all that life brought. I had to do it my way, struggling through brambles, up and over cliffs.

There are many times I have thought, "Kirpal Singh, if I were only more obedient, if I were only less curious, how blissed out I could be. I could be saying your mantra this very minute in an ashram in Vermont with nothing more to worry about than whether to keep the snails off the tomatoes with eggshells or rows of marigolds. But, no, I had a better way. I wanted something more exciting to worry about. I wanted to see if my back would make it another year; if I would ever meet my perfect mate and husband. But I am glad I knew you. I know we'll meet again. Thank you for coming to me and for your prayers."

I have lost track of Jan, too. I hear she lives in Canada with her husband and child. I read about her in a tabloid while waiting in line at a Safeway store.

Lola

Jan had a sister who was married to a man who used to be the husband of a wild woman named Lola, who took people to heaven through her facials. Not only did Lola's technique feel divine, it also lifted the muscles in the face for three days causing people to look ten years younger in a single treatment. Jan said Lola was anxious to quit the facial business and move to India to live with her guru. She was looking for someone to train for $1,000 in exchange for her knowledge and clientele. The deal sounded good to me.

Lola lived in an apartment on the beach at Malibu. When I went to her house, I could not get over the sound of the water. The Pacific Ocean was her front yard and pounded like some ecstatic, relentless god against the underpinnings of her apartment.

Lola was striking to look at. In her fifties, she was tall and blond. She had a laugh that erupted without the slightest provocation, as if she were hearing something no one else could.

There were many strange and marvelous experiences awaiting me when I entered Lola's world. Many left me a bit unnerved. First was the pink glitter she sprinkled outside her apartment entrance. In fact, Lola's whole place was pink. Her towels, the tile, her bed spread, even the pictures on the wall were pink. I'm surprised she had not sprayed the cat.

Lola had publicity pictures of all her clients arranged on one wall. For those who didn't have publicity pictures, the housewives or people not in the "biz," she would take their picture herself and hang them on her wall. If they were her clients, they were celebrities.

I studied her facial techniques for several months. To begin she would put fairy music on her tape machine and hot towels bubbling in the rice cooker. With waves crashing and shaking the walls, she would work on my face. Then I would work on hers, bringing fresh hot blood up the carotid artery into the neck, cheeks and forehead. There the blood fed all the cells while coursing through the veins on its way back to the heart.

One day Lola called me for a massage. I arrived at her home to find her high on acid, which, I learned, she took every month, supervised by a woman who came from Marin county. I had never heard of a supervised acid trip, but Lola did it for therapy. Lola laughed and shuddered and wept and screamed at the top of her lungs throughout the whole massage. At the end of the hour I was shaking. Afterward she leapt off the table as though nothing was the least bit out of the ordinary and handed me a check for $100. In the corner it said, "For whatever Lola wants."

She gave my name to all her clients. One woman had just wrecked her Rolls Royce while drunk. So when I went to massage her, she could hardly move. I worked on her shoulders, back and feet–the parts of them that weren't bruised. As I worked on her, I told her the story of how I got sober. She was so sick she couldn't talk, so I wasn't sure what she heard. Before I left she hugged me with tears in her eyes. I began seeing her three times a week. I massaged her for the next seven years. She moved out of town, but we still send cards and notes to each other and to my knowledge, she is still sober.

Another client referred by Lola was a beautiful woman pregnant with her third child. Toni was married to Dan, a man who built skyscrapers. I massaged this couple for nine years, every Wednesday night from eight until eleven. I don't know which of them I loved more. Dan was a riot. The first few years I massaged him, he couldn't shut up for a minute. He'd rattle on about everything, asking me questions, trying to pretend he wasn't nervous as hell about being in the nude with a woman other than Toni. He loved the way massage felt, but I could tell it made him nervous.

Getting Dan on the table without offending his dignity or invading his privacy was a comedy. I would pretend to need the bathroom so he could get on the table without me seeing him naked. Later I learned to hold the towel so he could undress behind it. Then he would lie down, and I'd drop the towel over him. Working these things out to the satisfaction of each client and their need for privacy is as much a part of the massage as anything.

Toni was just the opposite. She could walk down the hall stark naked, yelling at the kids, trying to create order out of chaos.

During the three hours I spent with them, there was a steady stream of children and animals. The children asked questions about homework, showed us pictures, tried to figure out what to wear the next day for school and discussed their teachers. Three dogs followed the children in and out and, at times, took up residence under or around the table, farting like crazy. The children always gave Hoover a bone outside on the porch.

I gave up trying to teach these people anything about relaxation and psychology and decided to just love them. They screamed and abused each other in a way I had never heard before. They also loved each other and stuck together through every kind of chaos. We laughed and joked a lot. Toni and I talked and complained a mile a minute through most of her hour and a half. Dan and I talked about more serious things. They said I kept them all from killing each other.

We often talked about Lola as it was Lola who got us together. Toni said she'd had an out-of-body experience while getting a facial from Lola. Those waves and the music lifted her right out of her body, and she was flying over the ocean before she knew it. While completely covered with steaming hot towels, she had also had visions of being stabbed. Lola could make her clients feel that the strangest things were entirely possible.

Just before Lola left for India, she told me she would write a letter to all her clients recommending me for facials. Later I found out she had given her clientele to her ex-husband's wife; the letter had never been written. I also discovered she had promised many people her clientele for $1,000. I was furious. I felt betrayed.

She returned home from India a month after she had left. India didn't suit her. It was too harsh. She was disillusioned.

When she called me, I asked her about our agreement and she apologized, saying she had needed the money and asked me not to be mad, that things had a way of working out. I didn't buy it. I said I thought what she did was dishonest and lacked integrity. She cried on the phone.

The following Sunday Jan called to tell me Lola had killed herself. I didn't think I was responsible, but I felt awful. I wished I had been more kind or forgiving.

Her suicide note said she had died laughing.

Although she didn't give me her facial clients, those she recommended for massage: the woman who wrecked her Rolls Royce, and Toni and Dan were enough. This incident helped me realize that life, the great rectifier, will give me what I need. I believe Lola knew this too.

Stars

My clients became more and more star-studded. My phone would ring, and it might be someone I had seen in the movies the night before or on television. Sometimes they were people whose names I'd never heard, but when they opened the door, I knew them at once. Because I was dyslexic and didn't read magazines, I would not know them until I saw their faces. Sometimes I would arrive at their gate and be met by a limo that would drive me down the mile-long driveway to the house, where I would be met by the butler. I entered these houses as though I were entering a museum. Nothing out of place. Everything polished and perfect.

The lives of these clients could not have been more unlike mine. In a single day I could massage a woman going to a party with the queen of England that night; or take a hot tub with, then massage, the wife of a man who was president of some oil company; or pound a few inches off the hips of someone going to the Academy Awards just hours away. After the massage I would go home to my house and listen to the roar of motorcycles ten feet from my windows.

I was not confident with these flashy people. I forgot my worth. I began to glory in the lives of my clients. These wealthy, famous people wanted me, and I lived for that. It was not unusual for them to give me $300 as a tip for a single massage. They would send their jets for me to come massage them when they were out of town, and give me front-row tickets to their plays. I wanted so much to be like them.

I absorbed my clients' attitudes and emotions. They came up my hands, into my arteries and washed through me. I stole their gestures and tried them on. I practiced being haughty, seductive, arrogant, and powerful. I carried their personalities with me for hours after a massage.

At times I would catch sight of myself and the life that had absorbed me. When I became aware of the changes in me, I realized how vulnerable I was. I likened myself to a piece of driftwood rushing downstream. I didn't know if I was being polished against the rocks or broken to bits. I didn't know until I entered calm waters and surveyed what was left.

Their lives reminded me of my childhood dream to be a singer. My parents were passionate about music—classical music. My father was president and my mother was secretary of the community concert series in our town. During the concert season, we had famous violinists, pianists and singers at our home. The pianists would come to practice on our large grand piano, and others would come to breakfast.

I was not allowed to go to the concerts when I was young. I could not hold still that long. My mother would dress up in wonderful dresses and high heels and go with Dad, leaving me with a sitter. I thought Mom was the most beautiful person in the whole world.

My father was a pianist, but he was never famous like the people who came to our house. When I was young, I vowed to become a famous singer for my father. But as my voice matured, it became apparent I didn't have a classical voice. I loved to sing the blues. When I heard the blues, I thought I heard my soul. But the blues were not allowed in our house, not even on my radio in my own room.

I sang in nightclubs from time to time, but I was frightened of the attention I received and stopped. I stopped singing completely except in the shower when no one else was in the house, or in the car on long drives. So I didn't become a singer. My dreams got buried and I lost my way. It was torture to be massaging people who were living out my dreams when I was not using my voice, not even to speak my mind. This went on for years; using these people for my identity because I had no true identity of my own. I couldn't be a star, but I could massage a star. I couldn't be a singer, but I could massage a singer. At least I could look into that world and imagine what it was like. I could touch it.

I felt like Sleeping Beauty surrounded by a thicket, silent as stone.

The Countess

One of my stars gave my name to a countess and her husband when they were "doing" L.A. one summer. He was a gangster from New Jersey and had leased a villa in Beverly Hills. They flew people in from all over the world to help them find "something to do out there in the provinces," and I massaged them all. The countess was used to life in the cultural centers of Europe, and she did not like L.A.

The gangster was very intrigued with Hitler and studied his methods as if in training for some dark position. He had been involved with governments and knew all about tortures used in Cuba, Vietnam and Korea. He had seen so much violence done to the human body that there were parts of him I couldn't touch. He would say, "We'll skip that part," and a shudder would go off through his muscles as a memory detonated deep inside his nervous system.

The countess was the gangster's whole life. She was a woman who could take the mind of the most tormented man and give him fresh, more immediate things to worry about. For example, she had a habit of disappearing. She kept the gangster chasing her around the world. I heard him on the phone one day after she had disappeared during the night. He paced back and forth across his room making frantic telephone calls to Vienna, Madrid, Paris or Hong Kong. She was a woman made for a man with a secret police and army at his disposal. She knew just what to do with a man like that. She did not take pity on him. She asked for what she wanted and got it. I worshipped her.

She had left heads of state and crown princes waiting at the altar, with newspapermen smiling at the door chewing on their pencils. I had married a man because he had cried when I said, "No."

"How could you do that?" I asked of one such incident. She, of course, thought it was their fault. "They did not have a sense of humor," she said. "They never would have understood me."

She was quite right. No one but the gangster could ever have understood her torment, her rage, her infidelities, her need to escape at a moment's notice. She needed to have someone track her down and bring her back to safety from the madness that drove her. "If he loves me, he will find me and forgive me," her actions seemed to say.

I believe it was this chase that tied them together. There were no more wars for him to practice his skills of espionage and no court to punish him for his crimes against humanity. She kept him vibrant and passionate, and at the same time miserable over her escapes, which soothed his guilt. By tracking her down no matter where she went, he proved himself and his love over and over again. Through these acts of devotion, intelligence and ability to forgive her anything, her admiration for him stayed alive, which is its own kind of love.

Other men had too much pride for such games. Not the gangster. Not the man who had done something secret in three wars. No, he had no pride. Or at least it wasn't based on something as insignificant as hurt feelings.

The gangster was a master of games of all sorts. I used

to play backgammon with him while waiting for the countess to finish her bath or while he was waiting for phone calls from his underground network after she had disappeared. I have never been beaten so quickly every single time. I accused him of having loaded dice and brought my own once. It made no difference.

The countess loved astrology, and once she did my chart. She was surprised I was a Scorpio. She said Scorpios were wild, intense people. I told her life had worn me down. She said that it was only a phase, and I would be deliciously horrid again. When her husband chided her about her belief in the stars she would say, "Go read something gruesome about Hitler."

The last time I saw the gangster, he was in despair over another of her disappearances. No army, guard or secret police could protect him from the devastation her leaving caused him. He could not find her anywhere. She finally sent him a note from a European hospital, and he rushed to her side. I heard he never left the hospital until, months later, she died of some stomach ailment.

He never called for a massage again. He never really liked them. He wasn't at home in his body.

I have thought so much about these two as their lives were inexorably woven into mine. Through loving and accepting the gangster, I healed the gangster in me. I have taken my own atrocities–leaving my son, betraying a lover, leaving the convent with the fisherman, breaking the heart of my mother, not remembering the names of God–and made peace with them.

Or perhaps it was the countess who healed them. She would have looked at my problems and flicked them away with a gesture becoming a queen and say, "You are a Scorpio. What can you expect from yourself? You were meant to break hearts. Break more. You have done nothing wrong. Forget it."

I have often wondered what the countess would say if she knew I used her voice, her haughtiness, and the memory of her gestures to exonerate my sins. I know that compared to the sins of the gangster and countess, my sins seemed insignificant. She would probably have a good laugh.

My friend Katherine went to a party one night at the gangster's villa after the countess had died. She told me he had made a shrine of her bedroom.

The Lamb

Frequently I exchanged massage with another masseuse named Kate. During one of these treatments she said she was going on vacation and asked if I would massage some of her clients while she was away. I agreed, and late one Sunday afternoon, I got a call from one of the men to whom she had given my name.

He was a handsome, twenty-nine-year-old president of something. His home was a mansion overlooking all of Beverly Hills and beyond. He had a manner of speaking that let me know he was used to having his way.

He was draped with a towel and I was massaging the knots in his neck when he told me what marvelous hands I had. This was not an unusual comment, but his tone of voice had changed. I recognized that voice. I had grown ears like microphones listening for that voice. It was the tone of Vaseline, of a person oiling his way into places he didn't belong. When I heard that voice my adrenaline kicked in.

To calm myself I talked about health. I quizzed him about his eating habits and told him clinically about his knots. He swore that he was in the best of health, that it was his work that made him tense.

When I got down to his legs, I knew what was coming;

he began to beg me to touch him. "Touch me, touch me, touch me."

We were alone in that large house. No one would have heard my scream. I kept telling him, "I don't do that kind of massage," all the time smiling, giving him my best work. I didn't dare make him mad. I was upset, but I couldn't show my anger. I kept effleuraging, petressauging and kneading. I think I began to hum. It was like saying, "Blah, blah, blah" when I was a child unable to cope with what I was hearing. I became that child again.

When the massage was over, I packed up my table without looking at him while he walked across the room to get his robe. He handed me my money, and I left. I cried all the way home in the car. Safe in my house I called a friend and told her what had happened. She said, "Why didn't you tell him to get the hell off your table?" I didn't know. I hadn't thought of it. Why couldn't I protect myself?

I confronted Kate about this man when she returned. She said, "I told him I didn't think you would go for it. In fact I made him promise to be good."

"What do you mean? Do you do sexual massage?" I asked astounded.

"Well, only when I want to," she replied. "It's very profitable."

I was shocked. I felt naive, provincial, stupid and curious all at the same time.

The incident with Kate's client stayed with me. I could not get him out of my mind. Why did I take everyone else's side against my own? Could I be all things to all people because I was no one in particular? Asking myself these questions and staying with the feelings, I finally got the answer. I had been six years old. I went across the street to play with my friend Bobby. Bobby wasn't home, but his uncle was. His uncle invited me in and took me upstairs to his bedroom. He laid me down on the bed. He said we would play a kissing game. He promised me a quarter a kiss for every place he kissed me. He kissed my neck, then he took off my clothes. Each time, before he kissed me, he asked, "Can I kiss you here?" And in my memory I heard a little, scared voice say, "Yes." I didn't know I could say, "No" to an adult.

Remembering that time, I entered the body of myself as a child. I swallowed my "no" and rationalized that I was going to get a quarter, and as I did so, I split my mind off from my experience; I disassociated. I betrayed myself with my own voice.

I couldn't tell my parents. I'd said yes to everything. When Bobby's uncle was through with me I wasn't even sure what had happened. The only thing I knew for sure was that he hadn't given me a single quarter, and I had been too shy to remind him.

After remembering the incident with Bobby's uncle, I was flooded with other memories, memories with names, dates and faces that before had only been screams or nightmares. I felt as if my brain was a large computer sorting through its files, flipping through its folds, muttering, "I know it's in here somewhere next to 'light,' no, 'lever,' no, 'lamb'–that's it." And lamb went with uncle, but not Bobby's uncle. Lamb went with my own uncle, Uncle Bill.

I never liked Uncle Bill or his wife, Myrtle. Bill always had tobacco juice running down his chin. He talked so slowly I would forget what he was saying. Myrtle talked fast, like the chickens. Even though Bill and Myrtle had chosen different styles of talking, both had taken on the sounds of the barnyard: caws, cackles and grunts.

Uncle Bill and Aunt Myrtle's farm always had kittens and puppies, hens and warm eggs in the hen house, and this particular Saturday there were lambs. I was shown the lambs and allowed to pet one of them. It was the softest thing I had ever touched. Then Uncle Bill told me to go away, and the lamb was led to the barn. When I tried to follow Uncle Bill and my father, I was told to leave, which made me want to stay all the more.

Uncle Bill tied a rope around the hind feet of the lamb and, with a pulley pulled it, upside down, into the air. The lamb bleated in terror and pain. I looked on in horror, then began screaming and begging them to stop, to let it down, and later, to hurry and put it out of its misery. I could feel the lamb's pain enter me. It was as if I were hanging there. Finally, my father demanded that I leave. I remember listening to the lamb bleating from the garden where I was told to stay. I could not stop crying.

Later I crept back to the barn and saw the lamb hanging with its throat cut and no skin. I knew my father and uncle had done that. I stared at the bloody body. I was filled with horror and sickness, trying to make sense of it. I dreamed of the lamb, imagining what its death must have been like. I imagined being tied upside down, my throat being cut, skin torn off, helpless.

Was this why I wanted to comfort the world with my touch? Was I trying to create a world of softness that I could live in?

The shock of remembering Bobby's uncle and the memory of the lamb cracked the wall that had been protecting me from my past. And as all my ghosts oozed out of the cracks, my emotions exploded out of me like a volcano, exposing all I wanted to keep hidden and burying all my defenses.

The Shadow Side

After my experience with Kate's client, and her response, the part of me that had been hiding behind layers of denial, temporary roles, New Age philosophies, prayers, yoga and abstinence came out and demanded its time and its voice. I knew I was in trouble the day a new client came to me. He was a scriptwriter who reminded me of Napoleon. When he lay on my table on his stomach and I worked at his head, he stroked my hips, and I didn't move away. After several weeks he said, "You have given me more pleasure than any one person in my whole life."

I found I wore less and less when Napoleon came. My thick, white trousers were replaced by thin skirts. My sexless white blouses were exchanged for tank tops or leotards. But the day I would have allowed him to take me, he was preoccupied with his movie. Something about the perception of his movie had changed for him. At one time he thought his script was a grand idea. Then on the day I wore no bra and no panties under my thin dress, on that day he saw his movie for the shallow, idiotic, boyish and trivial thing it was, and it took all the fire and confidence out of him, and I was spared. Though he came back for massage, the spell was broken, and my attraction to him died.

Next there was the businessman who came to my house to have privacy from his wife. It turned out that his wife and he had never been sexually compatible, even with four children, he said. He was confident of his good looks, though he did not appeal to me. I do not like pretty men. Give me someone with a scar or someone who has had his heart torn out. Someone who is lost or broken.

When I looked at this man, I saw the flabbiness of his character, and I hated him. He had the best of life too easily. He was a tree who had all the room it needed to grow straight and tall, the right amount of water, rich soil and sunlight. He had become proud of his thick branches and was stingy and vain.

And I was scared. My need for money, love and approval was struggling with my image of the "good girl." Two sides were warring within me, wanting to live a life of sex with strangers, wanting to tell this man how much I despised him, but fighting against the part of me wanting to take him in my mouth and be his darling. I wanted to shock him, tie his hands behind his back and tease every one of his pressure points. I wanted to arouse him until he came violently with a single kiss; I wanted to dive into the riptide that was pulling me under and down.

The day he called to say he was coming over, not for massage but just to see me, I wrote him a note and left it outside on my door. It read, "I don't want you for a friend, and I don't want you for a client. I don't want you to call me. I never want to see you again." Then I went to the beach and cried and walked Hoover until I knew the man had come and gone.

After that there was an actor who had been flirting with me for a year. He called from New York one day and asked if I would come to work on him before his play. I was flattered. He sent me plane fare, and when I arrived, he met my plane. He even took me to meet his parents. The play was a huge success, and I had a seat in the front row. He was attentive at the party afterward and introduced me to everyone.

After making love with him all night and a day, I massaged him with my hair and my thighs and my breasts. And as I sank deeper into this land filled with taboo, I realized that my fascination with sexual power and the thrill of the forbidden were stronger than my will to resist.

The confusion caused by having sex with a client—one I knew would never be my boyfriend or my husband—and my inability to stop it were a horror. It was as if the meaning of my life was just out of reach, and if I went just a little deeper into the unknown, I would find out what that meaning was. I judged myself horribly, the pain of which made me rationalize this behavior. "This was not a client," I told myself. "He was my client, then he was my lover, then massage and sex got confused. They were happening at the same time. That was not the same thing as being a whore, not the same thing at all." My guilt forced me to try to make my relationship with this actor more than it was, and we never saw each other again.

And finally there was Richard. Richard, the same name as my father. Richard, the same age as my father. Richard, who looked just like my father.

Richard's wife had heard about me and asked me to come massage them both. I massaged her first. Then he came bouncing in entirely naked. He hardly spoke during the first massage.

During the second massage, his wife was being interviewed by a women's magazine in the living room while I massaged Richard in the bedroom. He called me his little philosopher, his little poetess. In order to give him some humility, I told him he had a tummy. He said his sexpot wife put all her energy into her career, and he couldn't even get laid. He was so adorable and honest that my heart went out to him, and I couldn't help reaching down and patting his cock. He gave me $400 for the massage, though I charged only $75.

A month had gone by when Richard's houseboy called and said Richard wanted a massage. It would only be him as his wife was out of town. At the last minute I decided not to go. Then Richard called me from a health spa just a few hours from L.A. and asked me to come there; I gave in.

Richard had instructed me to wait outside the entrance to the spa. It had taken me two hours to drive there. I parked, got some coffee and drank it in the car while watching for Richard in the rearview mirror. Glamorous people walked by talking about glamorous things. Dressed-to-the-teeth women got out of Mercedes and Stingrays, Porsches and Rolls, and I became more and more insecure. I could hear the voices in my head begin to sneer. After forty-five minutes waiting in the red zone, I left a message that I had gone home. I had driven two hours and waited forty-five minutes to be a whore to a man who never showed up.

Part of me was thanking God as I drove away, and another was shaking with anger. I became hysterical in the car, driving down the freeway at seventy miles per hour. I was shocked when I began to scream, "Daddy, daddy, daddy."

I arrived home and went straight to the bathtub. I was going to lock myself in and not answer the door. I was going to call the six people coming to dinner that night and cancel. I was going to sit and demand that God change my life before I went on with it another step.

But the tide would not let go of me. The phone kept

ringing. I knew it was Richard. Finally I answered it.

He said, "Why did you leave?"

I said, "How could you treat me this way?"

He said, "Don't be mad. I told you I might be late. I couldn't get away. I was desperate to get away. I was so anxious at lunch that my family said I seemed preoccupied. I was looking forward to seeing you, and I am so sorry this happened. Please come back. You can spend the night."

"No."

"I will send my plane."

"Not tonight, I'm too upset."

"Well, tomorrow. I'll send my plane tomorrow."

I said I would see how I felt in the morning.

The next morning he called. He was coming to L.A., and I should come to his house at six. We would then fly to the spa.

When I arrived at Richard's house, all the usually locked doors were open. The pool was lit. As soon as he saw me, he raced over and began kissing me, unzipping me, pushing me toward the bedroom. I didn't expect this to happen until after eating and talking and flying and looking and flirting and kissing and touching–gentle touching. I did not expect this, not like this. No, no, not like this.

But I was not this man's woman; I was there for a purpose. We had agreed not to name it. We had agreed to put a mask on it. And all the while I remembered Kate saying, "It takes me days to recover. I am exhausted after. Your will is stripped away from you. Nothing is left. All your water is taken, and it comes back alone, in horror, drop by drop."

I thought these things lying on Richard's bed. I was not in my body. I was somewhere else watching, waiting for it to be over–until I got aroused myself and it brought me back, and I couldn't escape. I was there wanting, wanting. I couldn't stop wanting. Desire was pulling me into this man's arms, my father's arms, my rapist's arms, my molester's arms. And the guilt for loving it was unbearable.

How does a person live when conquered by an enemy? A body, mind and soul shackled by the shadow side–how does it survive? Words of war come to mind: internment, captivity, broken, demoralized, prisoner of war, dungeons, chains, slaughter, casualty, the dead, the fallen, the lost-in-action, numbered with the dead, massacred, war-torn. A person occupied by the enemy waits, listens and stays alive by not thinking too much, by barely breathing.

The first night I was with Richard, after sex and after dinner, we drove to the airport where we boarded his plane. An army of people were waiting for him. He told me, "Just smile and let them think what they want." Once in the air his steward brought us coffee and kiwi, and I looked out the window at all the bright lights.

The experience with him that night shook me to my core. I went places with him that rocked every idea I had of good and evil, right or holy. All those concepts were ground to sand, then melted. By the time Richard slept, my body was numb. I was thinking I might cry, but I didn't. My feelings were undistinguishable. I was emptiness itself.

That night I had a dream. Someone was making love to me. In the dream I opened my eyes and found it was my son. I woke up. It was as though my inner child was coming with a message. I did not understand. I felt sick.

At 9:15 A.M. as I was getting ready for the steward to pick me up from the spa in San Diego and take me back home to L.A., Richard followed me around showing me a birthday card he had picked out for his granddaughter and a new brochure for one of his stores. At the car he said he would call me when he got back to L.A., and then he handed me an envelope.

In the plane the bright sun made everything flat. The glittering lights of the city were burned out. I didn't look inside the envelope until I reached my car. Inside were ten hundred dollar bills. I sat looking at them, wondering who I was. All the way home and for many days after, I had a compulsion to buy ridiculous things I didn't need. I wanted to get rid of that money.

I knew I could not be Richard's mistress. I was not made to be a mistress. I would have fallen in love with him, and it would have been the end of me. I could picture myself following him around the world, sitting in suites in expensive hotels, waiting and crying. I would have had to be a secret. I knew I couldn't bear it no matter how much money Richard showered on me.

Then a miracle occurred. My son came to live with me. After thirteen long years living apart, except for summers, he finally came to live with me. I did not want to be a mistress. I wanted to be a mother.

When my son came to live with me, he was sixteen. It was 1985. He moved in with me because I lived in a school district that had a highly rated swim coach who said he could get Russell on a college swim team. Phil wanted Russell to have that experience.

When our living together became permanent, Russell's anger at me for leaving him began to emerge. He wouldn't talk to me. He would come home from school, go into his room and shut the door, coming out only for meals. He wouldn't tell me anything about his life, what he felt, what he was doing, what he wanted–nothing.

I sat in the living room hoping for a glimpse of him. To find out how he was doing, I eavesdropped when he called his father. My anxiety was almost uncontainable. I waited and waited–taking him to swim meets, praying while I cooked that the food would not burn, (I wasn't at all domestic), shopping for his favorite foods: chicken without bones, pineapple in chunks with small curd cottage cheese, processed whole wheat bread, spaghetti with canned sauce, no onions, oatmeal or cream of wheat with raisins and honey.

He would not celebrate Christian holidays, so we celebrated Jewish ones. Over Easter we ate peanut butter sandwiches on the beach and watched the surfers at Malibu Point while Hoover chased seagulls.

When Russell came to live with me, all my sixteen-year-old issues were coming to the surface. We were both emotionally sixteen at the same time; only I was in a thirty-nine-year-old body. But the mother in me was also longing to be born.

I believe it was through daily rituals that Russell began to trust me. It was small things at first: buying small curd cottage cheese, giving him a shoulder massage just before going to bed, watching M*A*S*H together on TV on a regular basis and picking him up at swimming practice on time. All these constant activities made it possible for him to open up and begin talking to me again. We began to heal the things that could be healed.

At the end Russell's senior year of high school, his swim team won the All City Championship. Mom, Phil and I went and yelled our heads off. There was a banquet to celebrate. Russell had been elected captain of the swim team that year and gave a speech at the banquet. He was eloquent. In the speech he thanked the swim team, and especially me, for making it the best year of his life. I could have died right then and felt my life was complete. After graduation Russell went to college in a town nearby and we saw each other often.

Even though I had stopped being Richard's mistress, and my son had come to live with me, my personal journey into darkness continued. Something was trying to come up; like a bubble of air from deep within a lake, or like a sliver finding its way out of the flesh. Something did not belong in me, and it was coming to the surface.

After Richard, all my attempts to find love ended in disaster. I could not look directly at a man. I stopped massaging for a while and worked as a secretary in an all girls' school. Then temptation came from women who said they wanted to have affairs with me. It was as if temptation came to me under the door, through the vents, through songs, through movies, plays and dreams.

And one day after lunch, driving down a street in Pasadena, my father said, "I wonder how we have lived all these years without making love? Do you ever think crazy things like that? I love you so much, much more than just a daughter. You are the love of my life."

I became numb when he told me this. I went into a survival mode, which was very familiar. All my feelings shut down as though a switch had been thrown. I could barely breathe. A smile appeared on my face. Three days later I hesitatingly asked a friend, "Is that a strange thing to say to a daughter?"

Nine months later when I told my father on the phone that I was seeing an old high school boyfriend who had found me after twenty-two years, he said drunkenly, "Does this mean I have lost you for good?" And he sobbed as if his heart would break.

A month later I wrote my father a letter. In it I told him these statements had made so much clear to me. I told him I always knew he had loved me inappropriately.

He wrote back that he had never said those things, that he had burned my letter, and that I should not ever mention it again. He couldn't imagine what I was talking about.

With each word of denial, I felt a bomb exploding. Shrapnel landed in huge chunks around me. Finally I opened my eyes, and the sun was gone and I did not know how to be a safe container for my feelings.

In time I began to understand why I never went to see my father, why I never thought of him, why he hardly existed in my life. In order to be close to him, I would have to tell him the truth about himself, and somewhere, long ago, when I had danced for my father, we had agreed to lie. An unspoken message had been conveyed: that his wanting me as a lover would never be acknowledged or admitted. It could be expressed through touch, lewd jokes, and glances, but we would never say the words. If it were put into words, the meaning would be known, and no one, not Dad or Mom or me could bear it.

When I accepted the truth about my father, other memories seeped back. I remembered that when I was a teenager he would get me drunk and come onto me as though I were a lover. Getting drunk allowed him to cross the line of taboos, and though I was disgusted I couldn't tell him. I protected him from knowing what he was doing. I protected him at the expense of protecting myself. I had been trained well by both my mother and my father not to expose them or even let them know what they were doing. It was as if I had been forced to become blind. And when he hugged me and touched my breasts, I never said a word. Not one.

And when I had my son–dear heart of my heart–I was terrified to touch him. His love reached in so deep. It went to the core of me where all my darkness lived–my atomic waste dump. Russell's sweet, little body terrified me. His helplessness was a torment. What was hidden in me could not be expressed. There were no words. I was never allowed to name my suffering. I could only run away from what I loved the most for fear of hurting that small person.

So some wise part of me had left him with his father until I got sober and knew what I was doing. Some good part of me, or some saint I had prayed to had mercy on us both.

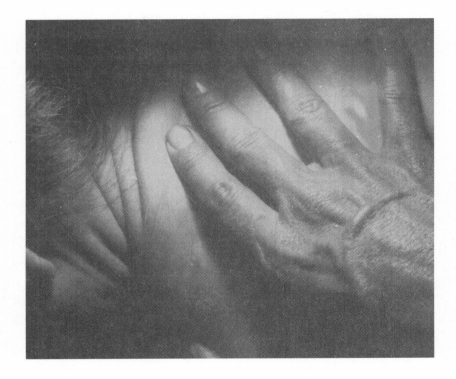

Memories Emerging

As I became accustomed to living with the feelings these memories produced in me, I began to realize how I had shut myself away. I had created a world I could live in. I had created a safe, protected room and lived in it. I comforted everyone who came to me as though they were my own self, the lamb, the victims of the world. Making friends with my rage, which I had never expressed, and with all the feelings of betrayal and loss, I began to attract women clients who had been through similar experiences.

Soon after my experience with Richard, I massaged Carla. We had been friends for a long time, but I had never massaged her. To begin, she lay on her stomach, and I rocked her gently back and forth a hundred times or more. As I rocked her the table creaked in rhythm like waves. As I rocked her, I remembered rocking my son, and myself, hearing lullabies, swinging, and leaves swaying.

Perhaps our friendship allowed us to move into a very deep state. I stood at her head, thumbs on each side of her spine. As I leaned forward they slid down her back. I hovered over the tension in her shoulders, her freeway-coffee, rush-filled shoulders. Her upper back was locked and stiff, holding in, holding back, but when I reached her buttocks I knew I'd found her holding flesh. She was so sore in her hips, I could barely touch her. Places that sore hold trauma.

As I massaged my friend's hips, she told me a story. She remembered it there on my table. "Remember" is not the right word because it had never been forgotten. No, she really let the story come out.

She had been six years old, playing in a barn, when a gang of older boys came in and raped her. She didn't tell a soul. For the next several years they hounded her, making her do it again and again. When she finally said "No, no more," the boys threatened her and then told everyone in the small, country town what she had done, that she had let them do it.

My friend told me that when she was in the second grade, she was afraid to look at anyone. They all knew what had happened, what she had done. When her father found out about the boys in the barn, he beat her. When her grandfather heard, he took advantage of the situation by molesting her himself, then warned her, "If you tell, I'll spank you."

Carla had been hanging her head in shame ever since, but she had forgotten why until the day I massaged her hips. Soon after that the rage came, and a very good therapist helped her remove those sick people she had been carrying within her for so long.

Though I never identified her, I carried Carla's story to a client whose name was Norma. I often use stories to help a client open up. Every story is a teaching. I was massaging Norma's chest, softening the muscles around the cavity where her lungs empty and fill, empty and fill.

Norma had asked me to give her an example of how the body stores memories so I told her about Carla.

During the story, Norma began to sob.

"What's wrong?" I asked.

Norma had a daughter, Martina. At three-years-old she went to a day care center. One day several boys entered the girls' bathroom. She was there, alone. They threw her to the floor, and while they held her arms down and her legs apart they raped her with the sticks and rocks they carried. They pulled the sticks in and out of her tiny vagina and when they were done, the boys walked away, leaving her there, all alone.

Martina didn't tell the teacher. Perhaps at three she didn't have language for what had happened. Perhaps she was ashamed. No one knew until her bath that night when her daddy lowered her into the warm water and she screamed in pain. Blood and pus were still visible. They rushed her to an emergency room where a doctor put her three-year-old feet in stirrups and, under the bright light, cleaned rocks and fragments of wood and dirt out of her.

Norma and her husband called the school and arranged to meet with the principal and the boys' parents. One family never showed up; the second couple got in a fight in the office, and the father said it must have been Martina's fault. The third couple said it couldn't have been their son.

Four years later when Martina entered regular school, she still sucked her thumb incessantly. She didn't pay attention, and she didn't participate. She had a rash around her mouth from nervously licking it. The teacher was frustrated and told Norma and her husband that Martina needed a psychologist.

"Did you tell the psychologist about her being raped?" I asked Norma.

"No, I have not thought about it for a long time," she said. "The story you told me brought it all back. How could I have forgotten?"

I was massaging Norma's feet. We were both crying. I had known this family for five years. Each member was precious to me. I was also crying because I knew how easy it was to forget. The mind wants to forget. After making breakfast a thousand times, going on vacations, getting dressed, doing laundry, taking the dogs to the vet, getting through the in-laws' visits, celebrating holidays, buying furniture, coping with deaths, going to movies, having financial problems, the mind covers up painful memories. We rage more and drink too much and can't concentrate as well, but we excuse it all by thinking, "That's what happens with age." We plaster over the memory of a baby girl lying on a bathroom floor being brutally raped. Instead of grieving we run marathons and shop more until that memory is sealed in solid iron.

The next day Norma told her daughter's therapist. He kept repeating during their conversation, "I can't tell you how important this is. I can't tell you how important this is. I can't tell you how important this is."

I called Carla, whose story I had told, to let her know a little girl was being helped because she had shared her story. "Don't be ashamed to tell your story," I told her. She said she would tell it from the street corner and sing it on the radio. Carla was forty-five before she remembered her own sexual abuse. So much of her life might have been different had her mother helped her deal with what had happened. When Carla finally told her mother, she told Carla of her own incest and how helpless she had felt. When Carla asked why she hadn't told her, her mother said, "You don't talk about such things."

And what happened to the boys? Where did they learn to hate women so young? Where are they now? Will they ever learn to respect their own softness or nurture a female body? Will they have baby girls of their own?

Creating A Cocoon

For weeks I grieved about Martina. Whenever I thought of her father lowering her into the warm bath, or her little feet in stirrups under the bright lights at the hospital, I fell apart. These images were unbearable to me. After hearing this story, I went home and went to bed. I took all that pain, both hers and mine, and felt the full power of it in my body. I held myself in that pain as though I were a hurt child. And I told myself, "I will not leave myself. I will not get up for anyone, until I feel like it."

Several days later, I was driving down Pacific Coast Highway when I felt all the blood drain from my face. My body felt as though I were waking up from a deep sleep. I remembered when I was sixteen and became pregnant by a boy I deeply loved. My parents nearly killed me. They put me in a hot bath for many hours and gave me pills my father got from a prostitute that were supposed to produce miscarriage. When that failed, they forced me to drink a pint of gin which was supposed to make me sick and cause the miscarriage. It didn't. I got alcohol poisoning instead. Then they drove me to Tijuana to an abortionist who tied me to a table. There was no anesthetic.

After that my mother watched for my periods like a hawk.

I paid dearly for my passion at the hands of my parents, but the offense that hardened me was of my own doing. The offense to my own life that chained me, like Prometheus to the rock, was committed months later after the abortion. My mother suspected I had seen my lover again, and she threatened me, "I will not go through another abortion with you. If you haven't started your period by tomorrow, I will put you into a foster home."

That night I went to bed with a snakebite kit and a Kotex pad. Mother would not send me away from my home, my school, the boy I loved. She did not know how scared I was. I thought I might be pregnant again. But I would not give into her again. I would not be vulnerable again.

I thought these things while getting into bed. Taking out the blade, I still didn't believe I would do it. Squeezing my hand, I made the blood rush down to the tips of my fingers. I still didn't believe I was capable of going through with it until some sort of trance came over me. The power to have my way went deep within me and connected with a source that had been there all the time. I just had to find it. It is a power that can be found only by blocking out the warning signals of fear, that listens to nothing that sounds like a whimper. This power has no pity, especially for itself. It is a power that abolishes the fear of God. It is hubris, pure and simple—the one sin the gods will not forgive.

I cut every one of my fingers and squeezed the blood out drop by drop onto the pad until the white cotton was stained bright red. I placed it between my legs and wore it into the right shape and showed it to my mother the next morning.

The last time there was a chance of being brought back to vulnerability and softness was a few days later. I was washing dishes with my mother, and she saw my hands. She picked up one hand and looked deeply into my eyes. There was horror in her eyes. She didn't understand; the meaning of what she was seeing would not come to her. I can't imagine she thought I cut myself like that by accident. She let go of my hand and walked away, not saying a word. She let go of my hand, and I was lost.

This power led me to having illicit affairs, drinking, stealing, doing drugs, dealing drugs, running red lights, sneaking out of places I belonged and into places I didn't. I didn't feel the consequences, didn't know there were consequences until I lost my son, and then, no matter what I did or where I put myself, the pain would not go away.

My mother eventually became exhausted from policing me and sent me to a foster home anyway. It wasn't so bad during the day. My life was filled with being in a new high school and finding my way through the maze of new faces. I was caught up in being sixteen, mastering my hair, my walk, my smile. It was here I learned to walk with my breath. When I walked to school, I walked slow, one breath for one step. I found this had a hypnotizing effect on me.

At night when everyone in my new home was peacefully asleep, I could not hold still in bed. I was so enraged I couldn't sleep. I just could not forget the abortion and the doctor telling me, "Don't move or the knife could slip, and you could bleed to death." I missed my boyfriend, Doug. My parents had the court issue an order making it illegal for him to see me until I was eighteen. I missed my cat, who had slept next to me, purring, for years. When my fury at being removed from my home and all that I loved overwhelmed me, I would creep out of bed and walk up and down the stairs, one breath per step.

One night the thoughts in my head would not stop. My mind kept screaming the names of my boyfriend, my cat, my horse. "Doug, Star, Mittens, Star. Doug, Mittens, Star. Doug, Mittens, Star." I would try to hold their images like a hug, as if naming them would bring them to me. But naming them only made their absence more real, and before I knew it, I was rushing over the edge, out of the lines, farther and farther out. I didn't think I could make it to the next moment if there wasn't someone to hold me, something to comfort me, something familiar. I began to say, "Wall, floor, ceiling, walking, carpet, stairs," pacing the stairwell naming the things I could recognize, trying desperately to have something make sense, make my mind safe.

I could feel the hysteria swelling in me, growing bigger and bigger inside me. I couldn't scream and wake the strangers I now lived with. How could I explain my scream to strangers? What would strangers do if my own parents, who were supposed to love me, who said they loved me, let me go, kicked me out? If they had done this to me, what would strangers do? I couldn't scream; I couldn't make a sound. I couldn't risk finding out what strangers would do. I could only open my mouth and pretend to scream. I could only scream in a whisper, "Wall, ceiling, carpet, stairs, wall, ceiling, carpet, stairs."

The pain in my chest became bigger than me and there was no way for me to control it, not by myself, not alone. Someone needed to be there, but there was no one, only strangers. And when it got too big to contain, I exploded. I flew apart into hundreds of pieces right there on the stairs. No one ever knew I exploded that night, but I did. I exploded into so many pieces I knew I would never be me again.

Exhausted, I went to bed.

I was amazed the next morning when Charlie Mae, the foster mother, recognized me. She made me breakfast and sent me off to school as if I were the same person I had been the day before. But I knew I wasn't. I walked slower than ever to school that day, not knowing whether I still breathed, whispering, "Flowers, lawn, house, red roof, doorway, tree."

And I heard someone inside me ask for comfort, and

very soon I was given a Marlboro. And I asked for comfort again, and the comforter came back Jack Daniels. And something rose out of me like a big black bird and flew through the land crying for oblivion. And all the dark things in the world heard and descended into me, feeding me, anesthetizing me, comforting me in my cocoon of emptiness, my cocoon of hardness.

Breakdown

Having remembered my abortion and that terrible night on the stairs, in the foster home, I began to crumble. I was standing in the kitchen wanting tea. I stared at the faucet and could not remember how to get hot water, the cup and the teabag together. The more I tried to push my mind to recall this magic, the more anxious and fearful I would get until I was hysterical. It was 1989. I was forty-one, not sixteen. Or was I? I was fragments of thoughts and images that couldn't come together. I was falling into unconnected pieces.

I would look at Hoover, whom I knew needed a walk. Then I would remember I hadn't eaten that day, and I would start bathwater. I would walk down the hall to the kitchen to get something to eat with one leg in a pair of jeans, the other bare for my bath and not be able to decide which should be done first or how to do it if I made a decision. I would end up pounding on my bed, screaming until I passed out from pure exhaustion, awaking the next day, every vertebra out of place. I screamed out the misery collected in my guts, veins, lymphatic cesspools, and the glands surrounding the pelvis that had captured my tears for years.

I couldn't hold still. I couldn't let the feelings come up. The clients who came, all with their rage or hunger or incest, began tearing away at the fabric of my cocoon. I emerged raw, without protection. It was my dark night of the soul.

In America we have no guides to explain this form of initiation. We call doctors who put us in hospitals and give us drugs so we won't feel as bad as we truly feel. They hide the Self from the self. When we buy the deceit and have successfully stuffed everything in place once more, the doctors sign our release, and we are free to roam the streets once more, slack-jawed with vacant eyes.

I did not take drugs. If I had, I might never have found my way out of the madness. Nor did I call a doctor. Instead I cried and pounded and ached and talked and wrote and cried and pounded and ached and wrote and ached and cried and pounded and ached and ached and wrote and cried and pounded and wrote for months.

I didn't dare go to the movies or listen to the radio. A strong emotional scene or song could send me into uncontrollable aching and longing that lasted for days. It took months before I had a glimmer of hope that I would ever be well again.

It has been my experience that if we become religious/spiritual without doing our psychological work, we often use the rules of a church or spiritual path or guru in lieu of maturing individually. For me to live authentically–that is, to be true to my own nature–I had to enter my heart. I did that by sinking down into my body, into my humanity, not trying to "rise above it." To live in a place of soul I had to sink down into all that pain, loneliness, despair, experience and rage and grieve until I burned out all that was false, leaving only my essence. I believe this is the true meaning of alchemy.

During this time, while all this was coming to the surface, I could not bear to see my mother. I did not see her for almost a year. When we finally got together, she was angry. She said I blamed her for everything. I screamed at her, "You almost killed me." I lost control completely as I told her about the long-reaching effects the abortion and the foster home had had on me.

After our fight I ran from the house and went down to the river. I thought, "She will never speak to me again." But within a few minutes she came to me and threw her arms around me. She said, "I am so sorry I hurt you. Please forgive me."

It was all I'd ever wanted, just to have her acknowledge what she had done. I had many friends and clients whose parents had chosen never to speak to them again rather than admit guilt.

My mother said, "Let's take two rocks and throw them in the river. They will represent our past." When I handed her a rock, she said, "That one's not big enough."

When we threw the rocks into the water, we agreed to start over, and we have. The empty place where love of mother should have been was soon filled with love again. We have become friends.

My awareness of the darkness in my life seemed to pull out the darkness in my clients. Everyone who came to me became aware of something that had been blocked. I believe that facing my own demons changed my chemistry and this change worked for others. I believe grieving is a political act. When we have the courage to grieve, we do so for everyone.

Sarah

Sarah sought me out because she had migraines and hoped massage would help. I learned she was obsessed with the homeless and did volunteer work to find ways to feed and house them. She always had a petition to be signed, a cause that needed funds.

I once went with Sarah on her rounds. She knew the names of the people I passed every day; people in front of the post office, sitting on the sidewalk amid a heap of rags and junk they had piled around them like a fort. I always thought of homeless people as mythic creatures or characters out of a Greek drama–some archetype that was coming out of the psyche of America that we had kept under and out of sight as long as we could. Sarah knew them by name.

Usually when Sarah had a migraine, she would call and cancel her massage, but one day the migraine came while she was on my table. Sarah said it felt like a monster, lurking in her shoulder, about to climb onto the base of her neck.

I said, "I would like to meet this monster. Can you ask it to talk to me?"

She thought my request was silly, but she agreed to play my game.

I asked the monster, "Who are you, and what do you

need?"

The monster replied, "I am the internal wounds, the deformity of Sarah. I want to be seen."

"By whom?" I asked.

"By her parents, the world, everyone."

I said, "I see you and you are enormous. I am very sorry you were so damaged. What do you need to feel better?"

The monster said, "I am unfixable. There is nothing to do. I only want to be seen."

I had to respect that. There are many things that are unfixable by anything we consciously know. Sometimes we need to include and make friends with the parts of ourselves that are unspeakably painful and horrible. I saw clearly why Sarah worked with the homeless; they looked like she felt inside. Their plight was enormously complex, their problems too large to imagine.

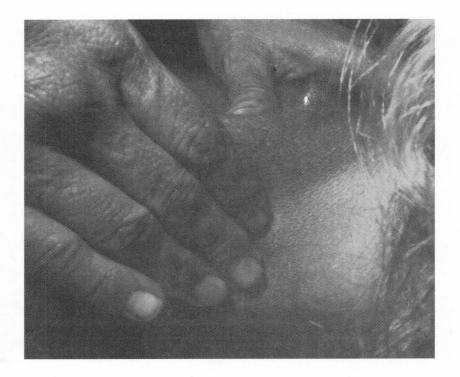

Katrina

When I begin a massage, the first thing I look for is the major spots of tension and pain. Then I ask, "What is the body saying?" The possibilities are endless. Shoulders can say, "Help me, I hate you, don't leave me." I have heard them.

I have heard feet say, "You are killing me; I can not go on." Backs, "It is hopeless." Necks, "Let go, let go." Arms, "Hold me."

Bodies can tell of witches in the pancreas, wolves in the spleen, and the need to spill your guts. We must listen and believe them.

I feel like a detective, staying open as each new clue presents itself. It is tempting to say, "Oh, this is why you are the way you are." The mind loves certainty, wants to know everything immediately, wants to be healed immediately, but pain is our guide. If we suppress the pain in our bodies, it will eventually find another way to express itself. It is telling us something we are unwilling or unable to say out loud, and so the body says it for us. If the pain could talk, it might say, "Hello, hello. You aren't saying how much pain you are in. You are smiling when you really want to hit someone."

Katrina was like that.

Katrina was the daughter of a judge, wife of a famous

journalist, mother of lawyers. She was an executive in a large corporation. I had never felt such all-over stiffness in my life, except for Mr. Pillsbury, who was dying of Parkinson's.

Katrina had had a car accident that left her with a whiplash, which is why she came to me. She reported almost driving into a pole on the freeway on her way to see me, and that weekend she'd almost driven off a cliff near Big Sur. I prayed for courage before I asked her, "Why are you trying to kill yourself?"

She was stunned by the question. She said she couldn't relate to the idea. She was very happy with her life. I knew she had everything she wanted and so this might be true.

"That may be," I said bravely, "but there is a part of you that is most definitely trying to kill you, and we'd better have a talk with that part of you."

Katrina was shaken by her near disaster. She would have tried anything.

I said a prayer, "Guide us safely." Then I told her to breathe deeply, then faster and faster. I had learned this technique, called rebirthing, many years before and had seldom used it. It brings up awareness from the unconscious so quickly that most people can't integrate it. Rebirthing opens the psyche by force. But I knew this woman and her grasp on reality. I knew it was life and death.

I heard her breath take over. She was no longer forcing herself to breathe; the breath became a force of its own. It was on the path, taking her back, bulldozing the walls that kept out memories, rushing through overgrown forests, and then a gasp. Ah, she is there. Tears down her cheek, yes, yes, she is there, in the room, in her father's closet the day of his funeral, smelling his jackets, saying, "I won't live without you. I want to die, too."

She had forgotten this vow, this promise sealed by emotion, but her body hadn't. That closet shocked her system.

I was shocked watching her.

Her vision was only the beginning. In her father's closet she turned around and said, "I will make earth my home. I will live out my life as it comes to me, no matter what the price."

There were years of integrating to do, pieces of herself to pick up, sort out, forgive. Ultimately she could choose to say good-bye to her father and promise to live with the living.

The Tear

Odette came to me after seeing an article of mine about stories in the body, published in *Women of Power*. She could not believe I had used the word "cunt" in print. She wanted to meet me because of that word. Amazing, our unknown signals reaching out to each other through the maze of media, through the channels of air filled with static and microwaves.

Odette was the youngest child and her father's favorite. At the age of four she was forced by her mother to make her father sign papers admitting him to Camerio Mental Hospital for depression. He had become intolerable to Odette's mother, and she was convinced that in the hospital he would get proper care.

Odette's mother piled all the children and her father in the car and drove to the hospital without telling anyone where they were going. Once there, her mother told Odette to plead with her father to sign the papers admitting him. Odette was not allowed to cry, nor ask about her father while he was "away." Her mother, who was overworked with six children, would not allow it.

Odette put her pain, her fear, and her love of her father in her back, neck, shoulders, hips, face, legs, arms. This perfect body, this body on which there was not one ounce of fat or cellulite, was rigid as stone.

Odette told me she had been in partnership with a man for three years. She said that she didn't love him, but she couldn't bear to hurt him by leaving. She couldn't leave her boss either, even though she hated her job. "What would he do without me? Who would answer the phone and pay the bills? Who would support him?"

When I said, "What were the words you wanted to say when you were four standing at the admitting desk at Camerio? Tell me your story of the silent ride home with your mother, brothers and sisters," Odette held her breath. A tear trickled down her cheek. Not a muscle in her face moved with emotion. Not a gasp of breath escaped her. The tear was the only sign she had heard me.

She never came back for massage. She called months later to say she would come back when she had enough money. I do not believe she lacked money. I think she was afraid of the feelings she now knew were inside her. To relax she would have to go through those feelings, and she just wasn't ready.

Mimi

I had known Mimi for years. At forty-five she was an executive in a company that did fundraising for organizations. She came to me for a massage when her regular therapist, with whom she had been doing incest work for years, was away on vacation.

Mimi told me that not long before she came to me, she had been praying on her knees in her bathroom, "Please God, give me the strength to face all that is blocking me from living completely." When she stood up, she lost her balance and hit her shoulder and head on the bathtub and lost consciousness.

While I massaged her neck, I asked her to blow out the color of the blockage she felt. Her neck was a mass of steel. The color of the block was black, she said. I felt so much resistance in me when I touched her. It felt evil. Not evil that was her, but evil that had come into her, perhaps while she was unconscious. Her color was gray: hair, skin, the feel of her–all gray.

I told her, "There is something very strong in you that is not wanting to leave you. It has made a home in you. It will not leave without a fight." She laughed and tried to make light of my remark.

After I had worked on her a while, I said, "I think we need

to call in some heavy-duty positive beings to counteract this beast."

I prayed for all the healing spirits to enter the room. I asked for the angels of compassion, abundance, faith, joy, kindness, forgiveness, freedom, love, gratitude, serenity, peace and right action. Some struggle was going on, but I didn't know how to grapple with it. It is so rare to find a being so "taken over." Someone can feel bad, even suicidal, but not be invaded. Knowing of this invasion frightened me. It told me that Mimi was not strong enough to ward off the evil that was present and always trying to find a home.

Mimi kept saying, "I'm so stupid. Why do I do these things?" I kept saying, "Mimi, you are doing your best. That was a strong and valiant prayer. Be kind to yourself. Love yourself." When Mimi left, I told her to treat herself as gently as she could.

That night I had a dream. There was a black, skinny creature with rat eyes, a goatee, a black hat and long, long fingernails. He took his fingernails, opened the skin of my neck and crawled inside my flesh. He lodged himself in the sinews there, in between the carotid artery and the tendons. I awoke instantly with stabbing pains that shot down my neck, shoulder and back. I realized I had gotten this creature out of Mimi and into me.

I immediately went to the kitchen and got an ice pack. I said, "Take that, you fucker." I knew I had to take action. Had I lain there like a victim, I believe this evil creature would have taken over my whole body. The creature hated the cold. I imagined him screaming, but I ignored him. At nine I called my chiropractor and made an appointment to have my neck adjusted. When I got home from the chiropractor, my son, now twenty-three, came over and gave me a massage. He was following in my foot steps. After Russell massaged my neck and shoulders, the pain was gone and has never been back.

I called Mimi to tell her what had happened. She was feeling better. I told her to keep praying. I told her to go to a doctor or chiropractor. She said she didn't have enough money.

Six months later Mimi died. The doctors say she died of ovarian cancer, but I know differently. I know she died of incest, loneliness, the inability to fight anymore, a lack of self-esteem that destroyed her ozone, and her immune system–that's what Mimi died of.

Perhaps she needed to start over with a new body, a new family. I pray she has a good rest. I pray she has an auspicious and beneficial rebirth, that she receives kind and loving parents who can give her the confidence and strength to flourish.

Unraveling The Mystery

I told all my celebrity clients that I was only massaging people who would come to my home. I could only be with people with whom I did not have to play a role. Sometimes I cried through the whole massage.

About this time I went to a therapist. Life seemed hopeless, worthless, just an endless unraveling of pain. My therapist, Jean-Claude, was the most joy-filled person I'd ever met. I told him I wanted to be free.

"Free from what?" he asked.

"Free from life," I replied.

"Well, if you were free from life, you would be dead. Do you want to be dead?" He was astonished.

My first homework assignment was to do anything I wanted to do. I went to a coffee shop and had cappuccino and a chocolate chip cookie. The first pleasure food I had allowed myself in a long time. I had become distrustful of pleasure.

By the second session I had decided to take a month off from massaging. I told everyone who called me about their incest issues or eating disorders or regrets or paranoia to find someone else to talk to–I was going on vacation.

By the third session I was a little scared about all the

space I'd made in my life. I was a plant in a new large pot—my roots were reaching out for the pot walls but there seemed to be no end. Something frightened me about all this space. My world had been so small before. There was a pain from my belly to my throat. Jean-Claude asked me to make the sound of the snake, rattling its rattles, over and over until I was as fierce as the rattler. My fierceness was not about saving my child or helping a friend or client. I was fierce for me.

I saw clearly that after I had shone my light down each dark hole of my psyche, grieved enough to make all the angels weep, and purged myself and cleansed myself and battled every dark beast the Greeks had ever dreamed up, I had to go on. Life insisted on it, or I would die.

My body knew it. I was bleeding all the time. My body had created fibroid tumors, which later were removed surgically. It is a common malady for those who have been sexually wounded. I thought of them as a symbol of my healing when they were removed. My body had arranged a way for me to die. If I wanted to live, there had to be an end to my grief or at least some balance with the joy of life.

Jean-Claude told me to do something outrageous every day. At first I couldn't think of anything remotely outrageous, but I soon got the hang of it. One morning I woke up and instead of getting on my knees and begging God for a decent day, I put on my tape of Gypsy Kings and woke up all the neighbors. During the day a woman related a sad story and instead of crying with her and asking how I could help, I said, "Death to all victims."

I had no idea what I'd do or say next. I dyed my hair bright red. I almost got a purple streak. Then I let my hair grow in gray. I stopped being so sweet, kind, and nice. When an old boyfriend gave me a peck on the cheek and introduced me to his new woman I walked away and didn't say a word. I wished them both dead (for several days) and prayed that she was nonorgasmic. Someone told me I was too angry. I told them, "If you can't handle it, leave."

I took up roller skating and African dancing and whipped myself into a Dionysian frenzy. I danced with my samba class in a parade. Once my lover climaxed first and I told him, "Honey, don't even think of sleep; you're not done yet."

I began to love my moles and wrinkles and accept my fear. I began to laugh with all my teeth showing. I cried with great huge sobs when I was hurt, and I began to tell my story, holding nothing back.

In time I was finally able to wake up from sleep without anxiety. I could meditate for hours on my breath without fear that the pain of being alive would kill me. I could look people in the eye, rich or famous, and not want a thing from them.

It was then I realized my breakdown was a breakthrough, and I began to relax. I began to feel safe with my feelings. The dignity of having survived the horror of that emotional and physical mutilation began replacing the shame, and I began to massage again. I now knew from experience that joy, bliss, happiness and pride could be attained after a childhood filled with "life-numbing" horrors. I consider it my good fortune to help people face, express and use the pain of the past in a creative way.

The people who needed what I had learned began to come to me. These hurt, abused, frightened people came for massage. They were the shattered parts of myself that had been lost and scattered long ago.

Claudia came to me after a car accident and was recommended by a chiropractor. Claudia talked of her own incest, her own loneliness, her inability to love. While she was coming to me, she got pregnant and had an abortion. Through her I once again relived my own. Together we wrote letters to our unknown, unborn children and asked their forgiveness for our not being able to be their mothers. Claudia and I held each other and cried, and it was bearable because we knew what we were grieving about. The cause of our tears had words.

Another woman came to me who was terrified of being touched but even more terrified of staying isolated. She walked into my office, belligerent as hell. She said a friend had made her come. She didn't want to be here. She would not take off her clothes.

After talking to her for a while I asked her, "What part of you may I touch?" After giving the question some thought, she said, "My feet. You may touch my feet."

And placing my hands on the feet of the woman who had not allowed anyone to touch her for years, I was overwhelmed with gratitude at the privilege. And as we worked together through her layers, the thrill and excitement of finding the person beneath the defenses almost could not be contained.

My massage room became a wild place at times. Calm songs were replaced with energizing rhythms. And when it was appropriate, I shook obnoxious rattles at all the demons. And I would place prehistoric mud on my clients' heart chakras and I had bones and crystals all over the place. We created all kinds of rituals to signify breaking out of our cocoons, which had kept us safe for all those years by being small and invisible. We danced for stallions in red boas and demanded back our passion. We wrote prayers on cloth, soaked them in menstrual blood, and burned them on the roof of my condo beneath a full moon. We prayed to wise elders on invisible planes.

Something changed in me and was reflected in my clients. The ones who now came were willing to look deeper within than they ever had before. I was a tuning fork. As I held a clear middle C, my clients struggled with the discordant events and traumas that kept them from their own true tones. The stories they brought me did not scare me now. I had been there and knew I could survive, no, not just "survive," but have a rich and meaningful life.

Now I often tell my clients, "Take one feeling and follow it down until it reaches through your pretense, all your politeness, down past the sweetness, past the place where you're just fine, down beyond where frustrations suck out your energy, where loneliness leads you to thoughts of suicide, to where your madness lies. Take that feeling and let it penetrate your madness like an arrow and send its poison gushing from you like a geyser.

"It is coming, your story. Stay in the silence until it comes, until something cracks. When the crack is large enough, you may hear a sound like horses galloping their way out of you. It is the memory of your story. The grief will come, then the rage for all that was taken: your hips and legs, your pleasure, the ability to trust or care deeply for another without terror. All that was blocking you from living is coming to the surface, and you welcome it with a war cry, much like the infant when first born, but this cry has power and wisdom."

As our bodies are unbound, our tension begins to melt. We no longer need arms, backs and shoulders like steel girders to protect us. We have become people who can say, "No" and mean it. Or say, "Yes" with all our hearts.

With each understanding, with each act of forgiveness I witness, with each act of loving a human transgression, I forgive myself at a deeper level. And I experience the truth that as we forgive each other, we ourselves are forgiven.

I no longer need anyone for my identity, though my clients give me a lot by wanting what I have to offer. I, in turn, give them a safe, warm place to unravel their mysteries.

It has been twenty years since I massaged the owner of the health spa. So many people are a part of my life. They want to come out and be seen and heard. I feel as if I hold their histories, their stories. At times I have tried to stop massaging and be something else. But then I hear a body calling my name and I am compelled to hear its story. I never meant to be a masseuse, but I suspect I always will be.

Epilogue

I like my life. I am proud to be me. What I have lived through has created me; the childhood events that formed me as well as the process that healed me. Both were necessary. I would not change anything.

Being a massage therapist and working with the bodies of hundreds of people, I have been in the presence of truth. The body can't lie. It can't say, "I don't feel bad," when it feels awful. Massage allows the body to stop and the mind to slow down and take stock of itself. It allows time to discover the constriction, pain, repressed memories, conflicts and emotions that one harbors in their body. Discovery allows choices. A person can continue to ignore their problems, holding onto them, letting them grow in intensity, until the person becomes sick or dysfunctional. Or, one can decide to acknowledge their problems, claim them, and begin the road to recovery and good health. The truth about one's life is in the body. Like a road map, that truth can be revealed through the co-creative process of massage and the willingness to hear the messages hidden deep within the body.

The more I am in touch with the source of my life, the more my clients are able to feel and acknowledge themselves. This means I must let go of all judgments, hopes and fears.

When my clients are able to relax and accept themselves and their lives, and when they feel forgiveness for all who hurt them and all the ways they hurt themselves, my massage room becomes a sacred temple. When this happens I, too, am transformed.

I can hardly believe what has become of my life. In 1995, after the completion of this book, I was asked to do a reading in Boulder, Colorado for International Women's Week. I was given an hour and a beautiful stage in the Old Main building at the University of Colorado. Instead of doing a reading, I created a one-woman show and performed it myself. It was very much like giving the owner of the spa a massage without ever having given one before. I am not an actress. I have not been on stage since high school, except to read a poem.

Terrified the night of the performance, I didn't think I would remember one line. I was pacing in my dressing room while the coordinators of International Women's Week introduced me. I thought, "I'll just peek through the curtain to see who's there." I just wanted to see what it looked like out there. When I left the dressing room, the door slammed shut and locked. There was no way out except through the stage. I thought I would walk out on stage and tell the audience to go home, but then I saw Michael, Raina, Odette, Jacqueline, Emily, Celia, and eighty others, and I could not walk out.

As soon as I began my performance, I became a person I had never known before. I call this persona "The Queen." She is all confidence, all knowing. She is a combination of Lola, the countess and Mrs. Adams, but she tops them all. I have gone on stage in every kind of mood, and she is always there.

I have given eight performances and received seven standing ovations. When I am receiving this praise, I often think of the girl I was and wonder how I got here. Now I am planning to tour with my play.

I am in awe at the twists and turns, and the mystery of life that make us first one thing, then another. Each day I give thanks for my life and the people who raised me. I have forgiven my parents for the hurt and pain I experienced while growing up. We now have a deep abiding love and respect for each other.

I am teaching a class entitled "Writing Your Memoirs" at a local senior center. I have written a manual called "Tell Me Your Stories," which includes assignments to help bring forth the stories people have lived. This class is fulfilling for me, and the people who have taken it say it has changed their life.

I still do massage. I cannot imagine not touching others in this way. It has become a meditation for me. It grounds me. I often give talks at the local massage school. It touches me deeply to see new massage therapists going out into the world seeking their fortune. I warn them: "Everything you do not know about yourself will be revealed through massage. Be compassionate, love yourself and all those you encounter. This is the real healing."